# Hydraulic Fracturing Explained

*GULF DRILLING SERIES*

*Casing and Liners for Drilling and Completion*

*Managed Pressure Drilling*

*Underbalanced Drilling: Limits and Extremes*

*Hydraulic Fracturing Explained*

# Hydraulic Fracturing Explained

***Evaluation, Implementation and Challenges***

Erle C. Donaldson, Waqi Alam and Nasrin Begum

Tetrahedron, Inc

Houston, TX

Hydraulic Fracturing Explained: Evaluation, Implementation and Challenges



Gulf Publishing Company
2 Greenway Plaza, Suite 1020
Houston, TX 77046

ISBN: 978-1-933762-40-1
10 9 8 7 6 5 4 3 2 1

**Library of Congress Cataloging-in-Publication Data**

Donaldson, Erle C.

Hydraulic fracturing explained : evaluation, implementation, and challenges / Erle C. Donaldson, Waqi Alam, and Nasrin Begum, Tetrahedron, Inc.

pages cm. – (Gulf drilling series)

Includes bibliographical references and index.

ISBN 1-933762-40-3 (978-1-933762-40-1 : alk. paper)

1. Oil wells–Hydraulic fracturing. I. Alam, Waqi. II. Begum, Nasrin. III. Title.

TN871.D584 2013

622'.338–dc23

2012047687

Printed and bound by CPI Group (UK) Ltd, Croydon, CR0 4YY
Transferred to digital print 2012

Printed on acid-free paper. ∞
Production services and design by TIPS Technical Publishing, Inc.

*Dedication*

*This book is dedicated to Dr. George V. Chilingar who has devoted his life's work to the education of many generations of students that have passed through the hall of the University of Southern California. Dr. Chilingar has also enhanced the knowledge of countless engineers and scientists around the world with a stream of fine technical books that have resulted from his untiring encouragement and assistance to countless co-authors of multiple technical books, and others that he authored and edited through the years. I am personally grateful for his guidance and patient advice throughout my own career.*

—Erle C. Donaldson

*This book is also dedicated to all the engineers and scientists who have contributed to the methods of development of energy that is sustainable and fuels our economic growth in a responsible manner. Dedication is also made to our parents, teachers, family members, and friends who inspired us in doing the right things in life.*

—Waqi Alam and Nasrin Begum

# Contents

# Foreword

The development of natural gas production from shale beds, which were previously by-passed, has expanded rapidly during the past 20 years. This development took place as a result of the improved technology of horizontal well drilling and hydraulic fracturing. Development of this new hydrocarbon resource in many areas that have not previously been impacted by oil production has increased the awareness of this activity with consequent questions regarding the technology, surface environmental impacts, and concerns of fresh water aquifer contamination by the practice of hydraulic fracturing. Hydraulic fracturing is noteworthy because of the relatively large amount of specialized equipment required for its implementation.

Shale gas is gradually increasing in importance as a natural hydrocarbon resource that has the potential for replacing a high percentage of the petroleum products currently in large demand. In fact, several large natural gas producing companies are contemplating the introduction of liquefied natural gas stations along the network of automotive express highways for use by vehicles equipped with engines that use natural gas. Thus natural gas may one day fulfill a large part of the demand for automotive energy and ease the future reliance on economically volatile imported hydrocarbon energy sources.

Fresh water aquifers near the surface are protected by state and federal regulations and refinements of technology. The accurate assessment of subsurface stress conditions and the mathematics of rock mechanics predict the size and extent of fractures with great accuracy. The use of micro-seismic monitoring during fracture propagation to precisely follow the actual growth and extent of the induced fractures in "real time" have all increased the precision of the technology. In addition, the practice of maintaining the integrity of overlying formations above the shale formation that is being exploited and the near-surface fresh water aquifers is an added safeguard.

At the end of the fracture stimulation (that uses as much as 2 to 4 million gallons of water) there is a period when a considerable amount (as much as 30%) of frac-fluid flows back to the surface when the well is placed on production. This fluid contains some of the special chemicals that make up the frac-fluid mixed with brine from the formation. This "flow-back" fracture fluid and formation brine is collected for proper disposal as authorized by state and federal regulations. In some cases the fluid is injected into a deep brine subsurface formation where it is permanently sequestered. If the flow-back can be recovered for reuse in another nearby fracture stimulation project, it will be collected in a tank and moved to the new site since this is most economical for the company. In any case, the water will be treated and processed for proper disposal or re-use; and a vigorous research program is on-going for development of new methods for treating water produced from shale gas operations for reuse.

This book is designed to explain the geological aspects (rock mechanics) of hydraulic fracturing in terms that can easily be understood, the technology of hydraulic fracturing and fluids used in the process, and the environmental concerns that have developed as part of the process. Mathematical concepts have been presented in their simplest form with careful attention to the explanation of the theories involved. Some theoretical issues have been removed from the text but are included as appendices for more comprehensive explanations. A few example calculations have also been incorporated as Appendix C for further analyses. A glossary of special terms has been attached and all acronyms used in the text are explained in the Nomenclature.

# Preface

The technology and applications of hydraulic fracturing have enjoyed tremendous growth following advancements of (1) horizontal directional drilling, (2) micro-seismic monitoring of drilling and fracture growth, (3) development of digital imaging software, and (4) the discovery of slick-water (addition of a polymer to fracture fluid that reduced the conductor tubing pressure loss of injected fluids). The first three blossomed together in the 1990s and underwent rapid refinement into the following decade when they were joined by the introduction of slick-water for fracturing gas-shale.

The developments of hydraulic fracturing technology were coupled to the excitement of the rapid discoveries of gas-shale around the world that could suddenly be produced economically and efficiently. This enormously complex endeavor burst into the public domain under a dark cloud of suspicion when allegations of careless, or inept, applications resulting in shallow fresh water contamination by fluids and gas began to appear in widely read publications. Part of the problem seemed to be a lack of understanding of the process: the use of chemical compounds for friction control, fluid viscosity enhancement to enable the conduct of proppants (generally graded sand), corrosion control, and the enormous amount of fluid that is required to fracture sections of long horizontal wells. Two fine, very technical, publications by the Society of Petroleum Engineers: *Reservoir Stimulation* and *Recent Advances in Hydraulic Fracturing* were available along with multiple papers of conferences specifically addressing hydraulic fracturing. All of these are readily available, but the problem associated with them is that they are written at a level that is understood only by engineers. Clearly, there is a need for a comprehensive text on the subject that can explain the salient technical aspects of the art in terms that can be readily understood by anyone who is truly interested in learning about the amazing technical advances that make hydraulic fracturing possible with fine controls that yield great technical precision in its application. That is the purpose and aim of this book.

## Acknowledgments

The authors would like to acknowledge the assistance received from Tetrahedron, Inc., specifically Mr. Daniel Ewald and Ms. Andrea Bouwkamp, in conducting literature search and drafting. Also, our gratitude goes to Mr. Robert A. Hefner IV for his encouragement and advice during the writing of this book.

# List of Figures

# List of Tables

# Nomenclature

## Acronyms/Abbreviations

| | |
|---|---|
| BHP | bottom hole pressure |
| BHT | bottom hole temperature |
| BMDL | bench mark dose level |
| CMC | critical micelle concentration |
| EDT | ethylene-diamine-tetra-acetic acid |
| EPA | United States Environmental Protection Agency |
| Ex. | example |
| Fig. | figure |
| GPM | gallons per minute (1 bbl/min = 42 gal/min) |
| HEC | hydroxyl-ethyl-cellulose |
| HPC | hydroxyl-propyl-cellulose |
| HPG | hydroxyl-propyl-guar |
| NMR | nuclear magnetic resonance |
| NTIS | National Technical Information Service, Washington, DC |
| RfD | reference dose for chronic oral exposure |
| SPE | Society of Petroleum Engineers |
| Tcf | trillion cubic feet |
| TDS | total dissolved solids |
| TOC | total organic content |

## Symbols

| | |
|---|---|
| $A$ | area |
| $A$ | empirical constant which is a function of the type of formation: sand, sandstone, carbonate |
| *Btu* | British thermal unit: heat energy required to raise one pound mass of water 1°F at one atmosphere of pressure |
| $C$ | constant associated with Poisson's ratio (usually equal to 1.91) |
| $C_b$ | bulk compressibility |
| $C_{eq}$ | equivalent conductivity of clay-exchange cations ($mho/cm^2/meq^{-1}$) |
| $C_o$ | overall conductivity of the formation ($1/R_t$) |
| $C_r$ | rock matrix compressibility |
| $C_w$ | water conductivity ($1/R_w$) |
| D | day |
| $E$ | Young's modulus of elasticity [($lbs/in^2$)/(in/in)] |
| $F$ | force (lbs-force) |
| $F_r$ | total, overall, resistivity of a subsurface formation |
| $F^*$ | corrected formation resistivity factor used in the Waxman-Smit equation |
| $G$ | shear modulus [$lbs/in^2$/radians] |
| $GR_{clay}$ | maximum gamma ray intensity in a zone of 100% shale, or value from laboratory measurement of a core |
| $GR_{cs}$ | gamma ray intensity recorded in clean sand |
| $GR_z$ | gamma ray recorded intensity at the zone of interest |
| $K_B$ | bulk modulus [($lbs/in^2$)/($in^3/in^3$)] |
| $L$ | length |
| $L_o$ | original length |
| lbsF | pounds force (lbsM/32.17 $ft/s^2$) |
| lbsM | pounds mass |
| $m$ | porosity exponent in Archie's equation, also known as the cementation exponent (default value = 2.0) |
| $n$ | saturation exponent in Archie's equation (default value = 2.0) |
| $P$ | pressure [$lbs/in^2$] |

| | |
|---|---|
| $P_{BH}$ | bottom hole pressure |
| $P_c$ | capillary pressure [lbs/in$^2$, Pascal], as defined in the text |
| $P_{cl}$ | fracture closure pressure, as defined in the text |
| $P_{col}$ | pressure exerted by a column of fluid in a well |
| $P_{fr}$ | pressure of the frac-fluid |
| $P_{fric}$ | pressure loss due to friction |
| $P_{LO}$ | pressure loss due to leak off of frac-fluid into the rock matrix |
| $P_m$ | borehole hydraulic pressure (mud pressure) |
| $P_p$ | pore fluid pressure of a rock sample or formation |
| $P_{prop}$ | fracture propagation pressure (bottom hole pressure) |
| $P_{res}$ | reservoir static pressure |
| *psi* | pressure: pounds-mass per square inch |
| $Q$ | injection rate (bbls/min) |
| $R$ | radius |
| $R_{sh}$ | overall resistivity of shale |
| $R_w$ | resistivity of the water saturating a subsurface formation |
| $R_t$ | total (overall) resistivity of a subsurface formation |
| $r$ | radial distance away from the center of a fracture initiation |
| $r_o$ | original radius |
| $\Delta r$ | radial distance from a seismic event to a geophone |
| $S_w$ | water saturation |
| $\Delta t$ | difference in time delay between the arrival of the P- and S- seismic waves |
| $u_c$ | compressive wave velocity |
| $u_s$ | shear wave velocity |
| $V$ | volume |
| $V_b$ | bulk volume |
| $V_o$ | original volume |
| $V_{sh}$ | volume percent of shale (should actually be $V_{clay}$ to represent the volume percent of clay in the shale which is furnishing the electrical conductivity) |
| $V_P$ | velocity of the seismic P-wave |

| | |
|---|---|
| $V_S$ | velocity of the seismic S-wave |
| $W$ | width of a fracture |

## Greek Letters

| | |
|---|---|
| $\alpha$ | poro-elastic constant |
| $\varepsilon$ | strain [in/in] |
| $\mu$ | viscosity |
| $\mu_a$ | apparent viscosity |
| $\nu$ | Poisson's ratio $\frac{du_x}{dy}$ (change of length with respect to change of width) |
| $\rho$ | density [g/cm$^3$] |
| $\rho_b$ | bulk density |
| $\rho_f$ | average formation pore fluid density |
| $\rho_m$ | formation matrix density (laboratory) |
| $\rho_z$ | recorded bulk density of the zone of interest |
| $\sigma$ | stress [lbs/in$^2$] |
| $\sigma_{ift}$ | interfacial tension [Neutron*10$^{-3}$/m] |
| $\sigma_d$ | stress applied at the outer diameter of a cylindrical rock sample |
| $\sigma_n$ | normal stress |
| $\sigma_r^*$ | effective radial stress |
| $\sigma_T$ | tensil stress |
| $\sigma_x, \sigma_y$ | horizontal stress |
| $\sigma_\theta^*$ | effective horizontal stress |
| $\sigma_z^*$ | vertical stress |
| $\sigma_\theta^*$ | effective tangential stress about the wellbore |
| $\Phi_d$ | porosity determined from the density log |

CHAPTER 1

# Hydraulic Fracturing Explained

## 1.1 Introduction

There is a lot of buzz going around regarding hydraulic fracturing—a process used to open new or existing cracks in the rock structures to produce oil and gas, also known as petroleum hydrocarbons. The cracks introduced in the rock act as channels to flow oil and gas from the rock into the well for production (see Figure 1–1). Hydraulic fracturing, also called "fracking," has been used for more than 50 years; however, its use has significantly increased over the last decade as we try to produce oil and gas from a new source called gas-shale. In the past, when the supply of oil and gas was abundant and the risk of interruption in the supply chain was small, prices of these commodities were relatively low and fracturing shale to produce oil and gas was not considered economically feasible. With the constant uncertainty of supply and an ever increasing demand for energy, triggered by higher standards of living throughout the world, the price of oil has increased significantly. The price of gas has also increased except in certain parts of the world that have an abundance of this natural resource. Whereas oil has a developed infrastructure for shipment from almost every corner of the world, gas is still used locally in most cases because of the high cost of shipment where pipelines do not exist. Therefore, the price is tied to local, or regional, markets rather than the global market.

The high demand for energy has initiated rigorous investigations and development of methods that can produce traditional and nontraditional energy cost effectively. The petroleum industry, which currently produces the largest percentage of the total energy consumed, has developed and improved several technologies that can produce

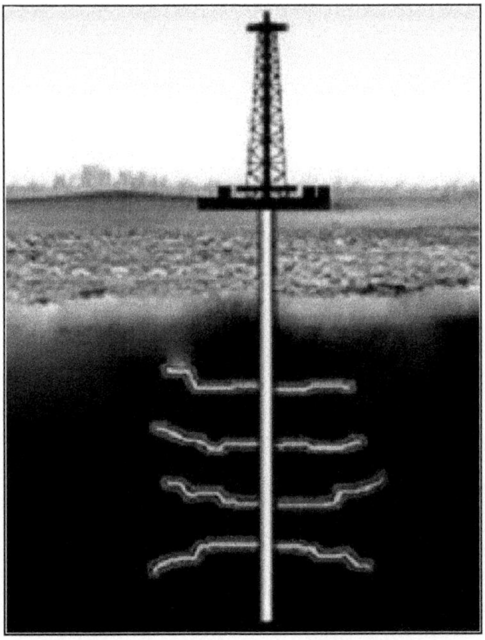

**Figure 1–1** *Fracturing the rock formation for production.*

greater quantities of oil and gas. The process of hydraulic fracturing has also undergone substantial improvement over the years and that has led to production of additional quantities of oil and gas. Unlike the past, hydraulic fracturing is now being conducted more extensively and sometimes near densely populated and environmentally sensitive areas and, as a result, has encountered greater challenges.

## 1.2 Petroleum Hydrocarbons

Petroleum hydrocarbons, consisting of crude oil and natural gas, have been used as a source of energy for thousands of years. They have been used as a source of heat and some consider them divine in nature when ignited because of the light and energy they emit. The Chinese are said to have used natural gas for hundreds of years before Christianity. The "eternal fires" that became the object of worship could have been caused by natural gas seepage from cracks in the ground ignited by lightning. It was not until the 19th century that natural gas was used commercially for the benefit of society. In 1821, William A. Hart drilled the first well in Fredonia, New York, to produce natural gas from the ground. This began the period when harnessing this useful source of energy began to look possible. It was during this time that street lamps were converted to natural gas, but homes were still not connected to the natural gas source. Towards the end of the 19th century and in the early 20th century, street lights started using electricity and the use of natural gas shifted to domestic purposes. However, until the 1970s, natural gas was considered a by-product of the petroleum industry, where most of the interest was geared towards crude oil, and great quantities of gas produced along with oil were burned, called flaring, in the oilfields, which became so bright that one could read a newspaper at night.

Crude oil production also started in the 19th century when Edwin L. Drake drilled his first well in Titusville, Pennsylvania, in 1859. Before Drake sank his first well, people around the world gathered oil for centuries around "seeps" (places where oil naturally rose to the surface and came out of the ground). Crude oil was used both as an energy source and also for medicinal purposes. Natural gas is often produced in association with crude oil, which in the past was mostly flared due to high transportation cost. With the depletion of oil reserves and higher demand for energy and synthetic products made from hydrocarbons, natural gas now has become a more important energy source. Its use and production has increased tremendously, facilitated primarily through advancements in drilling and production technologies, transportation technologies, and environmental concerns of alternative sources of energy.

### 1.2.1 Origin of Petroleum Hydrocarbons

Petroleum hydrocarbons exist in pore spaces of rocks in the subsurface, like water in a sponge. The primary difference is that rock cannot be squeezed to extract hydrocarbons. Heavier components of the hydrocarbon that are liquid are referred to as "crude oil" or "oil" while the lighter components are referred to as "natural gas" or "gas." Petroleum is a complex mixture containing thousands of different compounds (Tiab and Donaldson, 2012). Its composition can vary depending upon the environment in which it was formed. It is composed primarily of carbon and hydrogen and may contain other impurities. Crude oil consists of compounds that generally have more than five carbon atoms and are liquid. Natural gas predominantly consists of methane (a molecule containing one atom of carbon and four atoms of hydrogen) and may or may not be associated with crude oil. It also may contain some heavier hydrocarbons such as ethane, butane, propane, and even some liquid hydrocarbons, but their percentage of the total composition is generally small and varies from place to place. Hydrocarbons are formed in nature by the decomposition of organic matter under favorable conditions of pressure and temperature over millions of years in sedimentary rocks, generally shale, called "source rock." The composition of the organic matter along with the conditions of pressure, temperature, and time determine if oil or gas is formed in the source rock. Shale is a tight (very low permeability) formation in which oil or gas can move, but only slowly. Permeability of a rock defines how easily fluid can flow through the pores of the rock. Over millions of years, some of the oil and/or gas have migrated from the source rock, generally shale, and accumulated in more permeable formations, such as sandstone or carbonate rock formations, called "reservoir rocks" (Gidley et al., 1989). Reservoirs have confining layers, also called "seals," of impermeable rocks surrounding it through which the oil and/or gas cannot escape or migrate any farther. Reservoir fluids separate out in the reservoir rock due to their density differences, with gas being the lightest on top followed by oil and then water (see Figure 1–2). Reservoirs are, generally, substantially more complex than the horizontal pancake type layers shown in the figure. It is relatively easy to produce hydrocarbons from reservoir rocks because of their moderate to high permeability. Most of the oil and gas in the past have been recovered from land-based reservoir rocks. With the depletion of these reservoirs and ever higher demands for oil and gas, new technologies and frontiers are being explored. One such technological development has been in the area of gas production from the shale formations (which is a source rock) with extremely low permeability and, in many places, contains substantial amounts of hydrocarbons. These hydrocarbons are commonly called "shale-oil" and "shale-gas." The technology also facilitates production from other tight rock formations and coal beds that are not shale but have very low natural permeability. The technological leap that allows us to tap into these

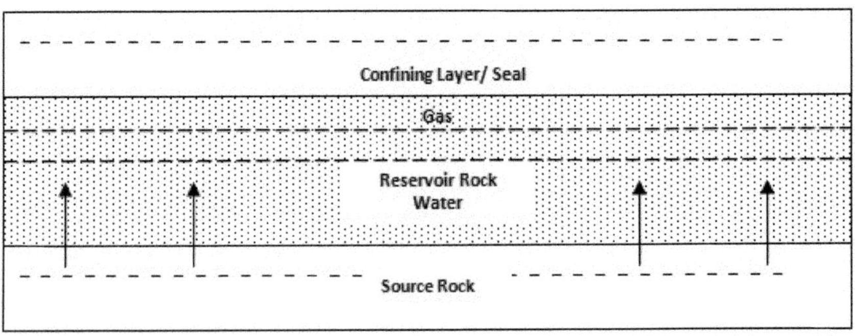


**Figure 1–2** *Origin and accumulation of petroleum hydrocarbon.*

resources of hydrocarbons consists of modern methods of directional drilling and fracturing. The directional drilling method provides the flexibility of constructing the well vertical, or at a slanting angle, to access the bulk of the hydrocarbon producing formation (see Figure 1–3). The fracturing process increases the permeability of the rock facilitating production of hydrocarbons, albeit with its own share of challenges, such as concerns with respect to the protection of groundwater, surface water, air, and farmlands.

## 1.3 Petroleum Reserves in Shale

The reserve estimate of the amount of petroleum hydrocarbons still in the ground constantly keeps changing because of new discoveries and depletion of existing resources. In shale, petroleum hydrocarbons are found in shale "plays," which are shale formations containing large quantities of oil and/or natural gas. These shale formations are geologically similar and are in the same geographical region. Estimation of reserves and extraction of oil and gas from these shale formations is still a challenge for the industry because of their complex nature. Production from these formations, using current technology, requires extensive hydraulic fracturing that allows the fluid to flow from the rock matrix into the production well.

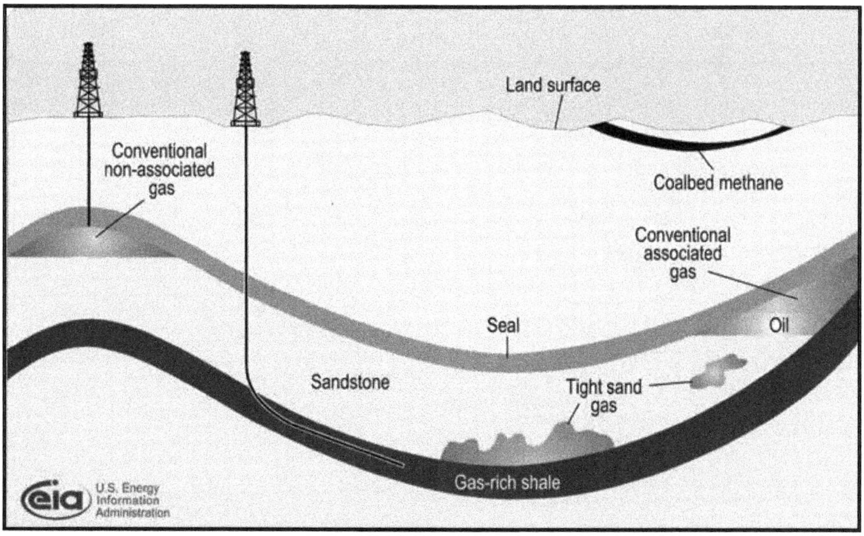


**Figure 1–3** *Production of hydrocarbons from geological formations using vertical or directional wells.*

Total proved world oil reserves (proved reserves are estimated quantities that analysis of geologic and engineering data demonstrates with reasonable certainty are recoverable under existing economic and operating conditions) are estimated to be a little over 1.3 trillion barrels (a barrel is equal to 42 gallons) (The US Energy Information Administration's International Energy Statistics, 2009). Shale oil is not a significant contributor to the total proved oil reserves. At the current consumption rate of approximately 75 million barrels per day (Journal of Petroleum Technology, May 2012), it should last us about 50 years. However, the consumption rate is expected to rise significantly.

According to the US Energy Information Administration's (EIA) Annual Energy Outlook 2012, estimated world recoverable reserve of shale gas is 6,600 trillion cubic feet (Tcf), which is more than one-third of the total conventional gas reserve. The estimate is based on data obtained for gas-shale formations (see Figure 1–4) around the world. Many areas that have the potential of adding substantially more reserves have not been studied. Access to this vast resource will require hydraulic fracturing unless some other technological breakthrough is achieved that is more cost effective and safe.

Table 1–1 shows that the potential of shale to meet world demand is significant. In some countries, such as France, where the estimated technically recoverable shale-gas reserve is 180 Tcf and where 98% of natural gas is currently imported, safe exploration and production of

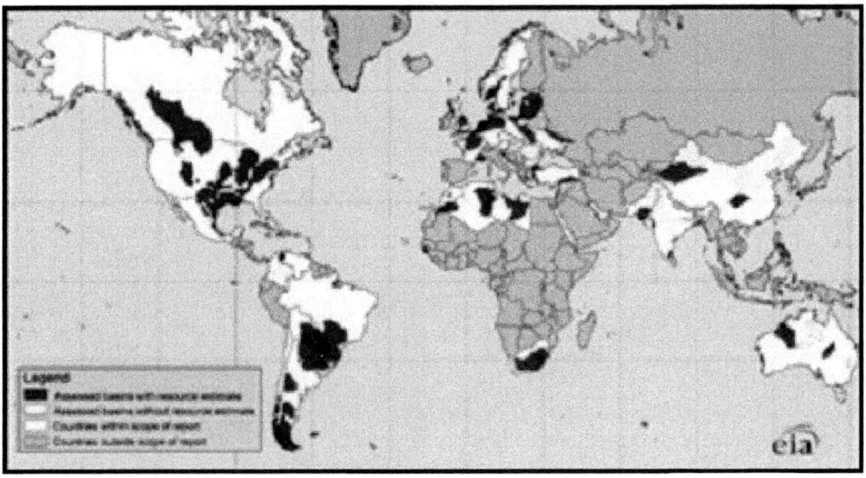

**Figure 1–4** *Cross-section view of the Eagle Ford Shale. (US Geological Survey 2010)*

shale-gas could be of tremendous benefit to the country. Situations in South Africa, Poland, and many other countries not listed in the Table 1–1, are similar.

The US Energy Information Administration's Annual Energy Outlook 2012 (US Dept. Energy, 2012) estimates that the United States possessed 2,214 trillion cubic feet of technically recoverable natural gas resources as of January 1, 2010. Of this, over 30% could be from gas-shale formations. Since hydrocarbon reserve evaluation of shale formations is still in its infancy, accurate estimates of resources are not available. At the 2010 rate of US consumption (about 24.1 Tcf per year), 2,214 Tcf of natural gas is enough to supply over 90 years of use. However, the rate of consumption will probably increase in the

**Table 1–1** World Shale-Gas Potential

| No. | Country | Technically Recoverable Shale-Gas Resource (Tcf) | Imports/Exports) Natural Gas (%) |
|---|---|---|---|
| 1 | Algeria | 231 | (183%) |
| 2 | Argentina | 774 | 4% |
| 3 | Australia | 396 | (52%) |
| 4 | Brazil | 226 | 45% |
| 5 | Canada | 388 | (87%) |
| 6 | China | 1275 | 5% |
| 7 | France | 180 | 98% |
| 8 | Libya | 290 | (54.7%) |
| 9 | Mexico | 681 | 12% |
| 10 | Poland | 187 | 64% |
| 11 | South Africa | 485 | 63% |
| 12 | United States | 862 | 10% |
| 13 | Others | 634 | |
| | Total | 6,609 | |

- Countries with significant amounts of shale gas and their current situation with respect to import or export of natural gas (Analysis & Productions, EIA)

future due to greater use of natural gas in power generation and automobiles because of its low cost and low environmental impact compared to other fossil fuels. Figure 1–5 shows the gas-shale plays in the United States. They extend across most of the country and some of the major plays are located near highly populated areas with sensitive environments in the surroundings. The Barnett shale in Texas, for example, is 5,000 square miles and provides 6% of US natural gas. The Marcellus shale extends across Pennsylvania, New York, Ohio, and West Virginia, and covers ten times the square miles of the Barnett, but it has only recently undergone incipient development (Am. Petrol. Inst., 2010). Technically recoverable natural gas (according to estimates published by the US Department of Energy) are: (1) the Marcellus shale with 262 Tcf, (2) the Haynesville shale with 251 Tcf, (3) the Barnett with 44 Tcf, and (4) the Fayettville with 41.6 Tcf (US Dept. of Energy, 2009).

Of the natural gas consumed in the United States in 2010, almost 90% was produced domestically; thus, the natural gas supply is not as dependent on foreign producers as the crude oil supply, and the delivery system is less subject to interruption. The availability of large

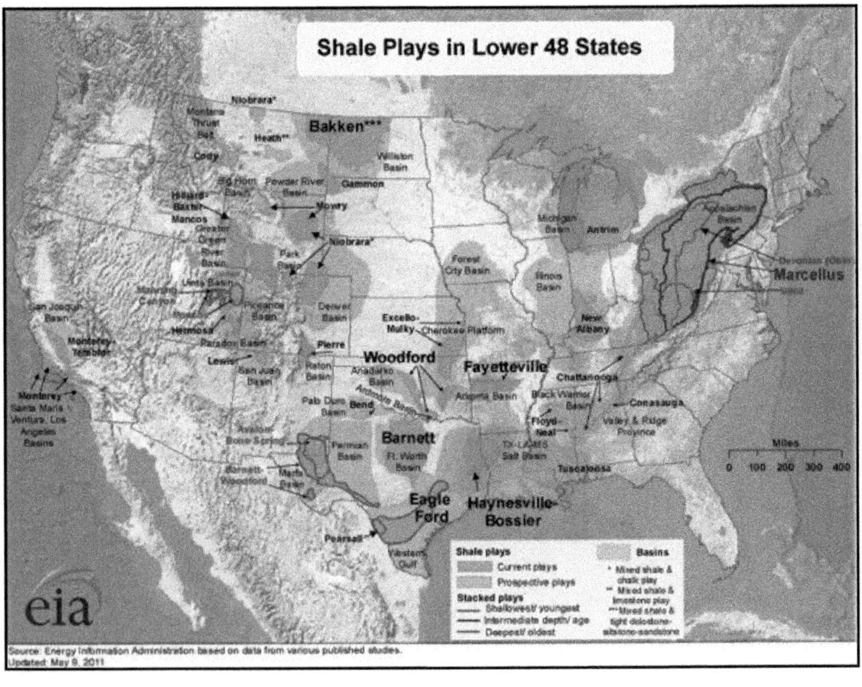


**Figure 1–5** *Major shale plays in the United States.*

quantities of shale-gas should enable the United States to consume a predominantly domestic supply of gas for many years and produce more natural gas than it consumes. Shale-gas is the largest contributor to the projected growth in production. By 2035, shale-gas production will probably account for 49% while tight, conventional, gas production accounts for 21% of US natural gas production, (see Figure 1–6). These production predictions are based on the use of available technologies including hydraulic fracturing.

## 1.4 Petroleum Demand

The world demand for energy has increased exponentially in the past and is expected to continue increasing for the foreseeable future, triggered primarily by population growth and higher standards of living around the world, especially in the developing world where most of the population lives. Energy is derived from two main sources: fossil fuels and non-fossil fuels. Fossil fuels include coal, oil, and natural gas while non-fossil fuels mainly consist of nuclear, hydroelectric, wind, biofuels, and solar energy. Most of the energy need is met by fossil fuels where coal is mainly used for the production of electricity. In the United States, fossil fuels supply about 85% of the nation's energy, of which about 62% is supplied by petroleum hydrocarbons, with oil supplying about 40% and natural gas supplying about 22% of the

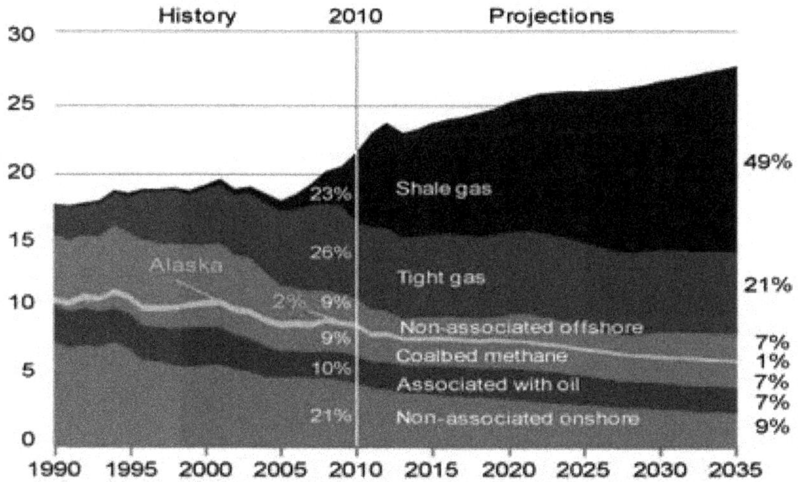


**Figure 1–6** *United States natural gas production, 1990–2035. (US Energy Information Adm., AE02012 Early Release Overview, Jan 23, 2012)*

total (US Dept. of Energy, 2008) as shown in Figure 1–7. With the increase in energy demands, we are observing a tremendous increase in the demand of fossil fuels as well, and this trend is expected to continue unless there is a major technological breakthrough that would allow safe production of energy in large quantities cost effectively in a non-conventional way. In addition to production of energy, fossil fuels are also used as raw materials for petrochemicals such as plastics, fertilizers, synthetic fibers, and medicines. They are the major source of hydrocarbons that are needed as the building block of our society as we know it.

The demand for oil and gas to meet the energy requirements has been rising ever since its discovery. Though both oil and gas provide good sources of energy, oil has been the focus of the industry for a long time. Until recently, natural gas was considered an unwanted by-product of the oil industry that was generally flared in the oilfield. Crude oil is easy to transport in trucks or through pipelines from remote areas into refineries or ports for shipment. Natural gas, on the other hand, requires pipelines or conversions to Liquefied Natural Gas (LNG) or Compressed Natural Gas (CNG) and is shipped in high pressured vessels. In addition, the energy content of natural gas with respect to its volume, under normal conditions, is less than oil. This is one of the reasons the industry had been focusing on oil, but with the depletion of oil reserves and environmental issues related to oil production and transportation (such as the incidents at the Deepwater Horizon/Macondo blowout in the Gulf of Mexico and Valdez oil spill in Alaska), focus is now shifting towards natural gas. Over the years, oil and gas that were easily accessible have mostly been depleted. We are left with petroleum reserves that require greater efforts to extract. The world still has substantial amounts of oil and gas left but they are more difficult to harness. Technological challenges facing the industry to produce oil and gas from these more complex situations require innovations and investments and must be produced in a manner that is protective of the environment.

Figure 1–8 shows the world demand for oil based on Energy Information Administration data of 2009 that has been extrapolated to the year 2030. The new economies of the world will push the demand higher and that will affect the total demand on oil. With the depletion of oil, this demand will have to be met by other sources of energy, such as natural gas.

Natural gas is extensively being used for the generation of power and its usage is on the rise because of its lower cost and lower environmental impacts compared to other fossil fuel operated power plants.

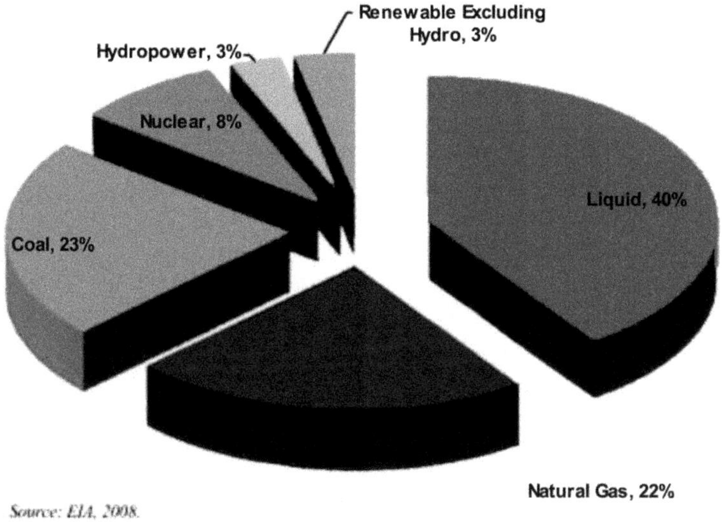


**Figure 1–7** *Source of energy in the United States. (EIA, 2008)*

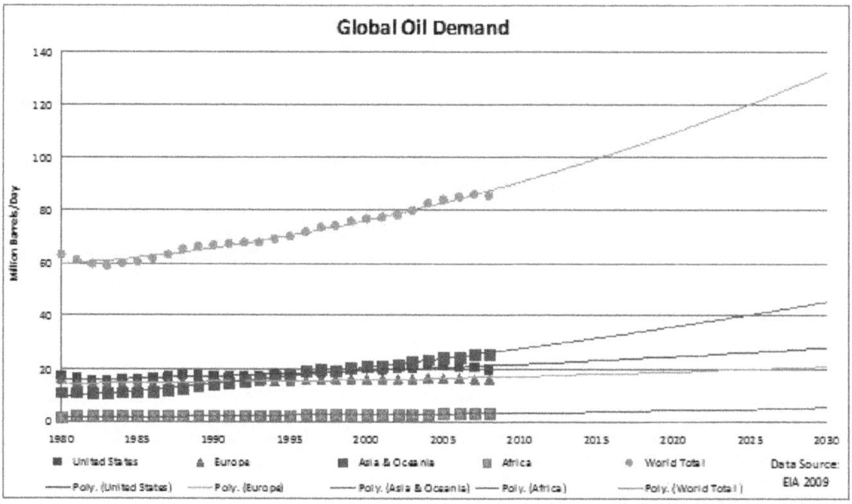


**Figure 1–8** *World demand for oil. (Extrapolated from EIA report on world demand for oil, 2009)*

## 1.5 Achieving Production of Hydrocarbons to Meet Demand

Petroleum hydrocarbons generally reside in complex geologic formations in the subsurface. They are often several miles below the surface of the ground in a porous rock matrix that acts as the reservoir. Some of the deeper reservoirs are more than 5 miles deep. Drilling down to these reservoirs is like drilling from the top of Mount Everest to the sea level. The reservoir is not a pool of hydrocarbon from which oil and gas can be extracted easily. Instead, the rock has a complex porous system that is similar to a sponge that cannot be squeezed and in which all pores may not be interconnected; thus making the flow of fluid difficult. The measure of difficulty or ease by which the fluid moves in the rock matrix is called permeability, and the percentage of the rock volume in which fluids can reside in the pore space is called the porosity. Rocks with higher porosity have higher capacity to hold petroleum hydrocarbon. In general, if the porosity is high, the permeability also is high. However, this phenomenon does not hold true all the time. One of the major factors that define the permeability of a reservoir rock is the nature of the interconnectivity of the pores in the rock matrix. Sometimes the rock formations may have large pore spaces where the petroleum hydrocarbon can reside in large volumes, but they may not be extractable because the pores are not sufficiently connected and, therefore, the permeability is low. This is typically the situation in shale formations. Shale has considerable porosity but low permeability because the sizes of the pores are very small. In order to produce fluids from these formations commercially, the permeability must be increased artificially.

In the previous section the known world reserves of oil and gas were discussed, but only about 30–40% of these resources can be recovered from conventional reservoirs of sandstone and carbonate using conventional technologies. Recovery from shale and other tight (very low permeability) rock formations is negligible. To increase production, various stimulation techniques are used in the industry. Some of them are described below.

### 1.5.1 Fracturing

Fracturing is a process of improving permeability of a tight rock formation such as shale to stimulate production of oil and gas. Fracturing is achieved by injecting a liquid into the rock matrix at pressures high enough to cause cracks or fractures to form. These fractures are created through the wellbore, and some extend for hundreds of feet into the formation. In many cases, very fine fractures (called micro-fractures) that exist naturally in a rock matrix but are closed due to cementation, salt deposits in these fine channels, and tectonic stress, can also be opened by injecting water solutions at high pressures. Once the fractures are created, they are kept open by injecting particles (called proppants) into the fractures; otherwise the fractures will close due to the normal subsurface stress in the rocks (the process is called healing, reverting the rock back to its natural form of low permeability). This is similar to opening a blocked artery in a human body and then placing a stent to keep the artery open so that the blood can continue to flow. Reservoir rocks are seldom uniform in their properties. Permeability within the same rock formation varies greatly. Flow of oil and gas into the producing well can be reduced drastically by a low permeability zone near the wellbore. The wellbore, from which the oil or gas is produced, can be considered as the bottleneck, and the reservoir as the main body of the bottle. No matter how easily fluid flows through the main body of the bottle, the eventual flow is controlled by the constriction at the bottleneck. If the bottleneck is clogged, there will be no flow from the reservoir irrespective of the properties of the rest of the reservoir. By fracturing the zone near the wellbore, the bottleneck is opened. Therefore, hydraulic fracturing may be required even in reasonably permeable reservoir rocks that have low permeability near the wellbore. To conduct hydraulic fracturing a mixture of water and sand, along with some other additives to produce specific effects that will be discussed later, is used. The liquid solution is injected at a high pressure and rate, which results in creation of fractures in the rock formation. Shale formations have micro-fractures that exist naturally but do not conduct fluid as they are not connected with a larger network of fracture channels. In addition, most of these micro-fracture channels are closed due to deposit of cementation material and natural rock stress. Application of fracture pressure and chemicals opens these micro-fractures and connects them to the induced fracture that facilitates production of oil or gas from the shale formation. Hydraulic fractures provide a conduit for fluids from the rock matrix to flow into the wellbore for production. It is estimated that over 90% of the wells now require hydraulic fracturing for production that is economically sustainable.

### 1.5.2 Acidizing

The formation surrounding the wellbore is treated with acid to increase permeability. Flow of oil and gas can be severely reduced, and even stopped, if the formation around the wellbore is clogged. Well acidizing is achieved by pumping acid (generally hydrochloric acid) into the well to dissolve limestone, dolomite, and calcite cement between the sediment grains of the reservoir rocks. Hydrofluoric acid also may be added to remove quartz, sand, and clay from the reservoir rocks. Acid fracturing is also conducted to increase permeability of a hydrocarbon producing formation. The process may be considered fracturing rather than acidizing.

### 1.5.3 Water Flooding

To produce incremental oil from a reservoir that is depleted or close to depletion, water is injected from one or several corners of the reservoir to push the oil into production wells. The system works well if the density, permeability, and viscosity of the oil is compatible with the injected water. Also, the reservoir rock characteristics play an important role in the success of water flooding. Several additives can be added to the water to make it compatible with the oil to increase production. The process is, generally, applicable to oil reservoirs and is not very successful if the permeability near the wellbore is very low.

### 1.5.4 Chemical Flooding

Chemicals such as surfactants and polymers are added to water to improve recovery of oil. Oil and water do not mix due to high interfacial tension between the two liquids. When water is injected in the reservoir, the water tends to bypass the oil. Surfactants assist in lowering the interfacial tension and producing the oil along with the water. This is similar to adding soap to greasy hands in order to release the grease.

The polymer increases the viscosity of the water solution (thickens the solution). Oil is generally viscous and therefore it does not move easily—it moves more like thick molasses. Water with increased viscosity is able to push the oil out of the rock formation. Like water flooding, the process is generally applicable to oil reservoirs and is not very successful if the permeability near the wellbore is low.

### 1.5.5 Carbon Dioxide Flooding

Carbon dioxide is injected in the reservoir to add pressure, especially in reservoirs that have lost pressure due to depletion of oil and gas. The added pressure helps produce additional oil and gas. Injection of carbon dioxide (which is a greenhouse gas) also helps reduce carbon dioxide from the atmosphere. Reservoir compatibility is studied before injection of carbon dioxide to ensure that the reservoir is not damaged by formation of carbonate salts that can reduce permeability. The process is not very successful if the permeability near the wellbore is very low.

### 1.5.6 Thermal Recovery

Thermal recovery is a process for recovering oil that is very viscous by applying heat to reduce the viscosity, facilitating the movement of the oil to the producing well. Fire-floods and steam-floods are two major thermal recovery processes. In the fire-flood, oil at one end of the reservoir is ignited and burns to create the necessary heat. The heat travels through the reservoir and reduces the viscosity of the oil. In steam-flooding, steam under high pressure is injected into a well to contact the oil and lower its viscosity. The process is applicable to oil reservoirs.

### 1.5.7 Microbial Enhanced Recovery

Microbes are added to the reservoir to produce additional oil. Various types of microbes produce bio-surfactants, biopolymers, and gas that help increase production. Microbes can also help improve permeability by removing wax from the vicinity of the wellbore. Since petroleum reservoirs are deep, there is no oxygen in the reservoir but there are microbes that can survive under such conditions, called anaerobes.

Of the various technologies discussed for stimulating production, hydraulic fracturing and acidizing are applied most widely. Without having reasonably good permeability, none of the other technologies are able to stimulate production.

## 1.6 Hydraulic Fracturing

The process of hydraulic fracturing requires water and sand with some additives. Historically, the process has consisted of non-water types of fluids, which have now been replaced by mostly water. Grebe and Stoesser (1935) were the first to propose the idea of using high-rate, high-pressure injection of a fluid into a well to create fractures extending away from the well in order to stimulate oil and gas production. Veach (1989) reported that hydraulic fracturing was first applied to a well in Grant County, Kansas, in 1947; about 1,000 gallons of napalm-thickened (or gelled) kerosene containing a suspension of 0.5 lbs of sand per gallon was injected into the formation at 2–5 bbl/min to create the fractures and prop them open with the sand. After the formation was fractured, about 2,000 gallons of kerosene containing a compound that would cause the gel to disintegrate into a solution was injected to remove the napalm gel that was used to create the fractures; some of the free-fluids were recovered when the well was placed on production. As they gained experience, the rates of injection increased to 500 bbl/min or greater, and the amount of fluid used for the initial fracturing stage was increased to more than one million gallons.

The process was offered commercially (under a license) by the Stanolind Oil and Gas Company (Clark, 1949), which continued to use gelled kerosene for the reservoir fracturing fluid. As news of the success of the process for stimulation of hydrocarbon production spread through the petroleum industry, hydraulic fracturing rapidly became a standard technique for enhancing oil production from wells that were declining in production, and for new wells that were drilled for injection of water to increase oil production (water flooding projects). Along with increased application, the technology has evolved as new discoveries and improvement of frac-fluids, methods of application, and surface equipment are employed and the science of hydraulic fracturing continues to develop.

The discovery of vast reserves of natural gas in the huge shale deposits increased the current estimates of recoverable natural gas in the United States to about two thousand trillion cubic feet. The produced gas is used for manufacturing, power generation, and residential heating, but production has reached the point where it exceeds the current demand for natural gas and now there are cutback by some of the large gas producing companies. Therefore, several of the large companies are considering the installation of natural gas stations along the expressways of the United States to facilitate the use of natural gas for automotive transportation; if natural gas were readily available, the demand for it would increase the demand for vehicles with engines converted to, or manufactured for, natural gas use. Natural gas is an excellent source of thermal energy because of the ease of transportation in pipelines and its thermal content: gas currently distributed to customers is blended to have 1,000 British Thermal Units (Btu) per cubic foot of gas (one Btu is the heat energy required to raise the temperature of one pound of water by 1°F at 1 atmosphere of pressure). The composition of natural gas varies from one location to another, but in general it contains ranges of low molecular weight gases as follows: methane >75%, ethane 10–15%, propane 10–15%, butane 10–15%, carbon dioxide 0–5%, and nitrogen 0–3%.

As the ability to drill deeper wells in search of hydrocarbon reserves increased, so did the troublesome problem of borehole collapse Expensive remediation work-overs and bottom-hole completions (re-drilling, cementing, installation of pipe casing, packing with gravel, etc.) caused by rock instabilities led to the study and application of the theories of tectonic forces extant in subsurface formations which were being developed by mining engineers (Hubbert and Willis, 1957; Cook, 1967). Research was then focused on cores from deep wells using laboratory equipment designed to simulate deep reservoir conditions of temperature, pressure, pore fluids, and three dimensional stresses (Donaldson et al., 1980).

The objective of fracturing an oil or gas formation is to increase the productivity of the well by forming fractures in the rock extending for some distance into the formation. The fracture increases the area for drainage of the hydrocarbons into highly conductive channels leading to the well. Thus, the hydrocarbons can be produced at greater rates and at lower pressure than is possible from an un-fractured well. The fractures will close under the normal reservoir horizontal stress extant in the formation just after the pressure of the fluid used to create the fracture is reduced to the normal formation pressure. Therefore, it is necessary to introduce some type of particles to hold the faces of the fracture apart and afford a highly permeable conductive channel from the formation to the well. The particles are called *propping agents,* or *proppants,* and consist of well-sorted sand, glass beads, walnut hulls, and many other types of synthetic particles. When an acid is used for the fracture fluid, the acid removes acid-soluble portions of the rock, leaving etched channels along the walls of the fracture that remain in place after the fracture closes. Thus, no propping agents are required if the formation can be etched by an acid frac-fluid.

As mentioned above, a hydraulic fracture is initiated adjacent to a productive formation from a perforated cased well, or an open-hole below the well casing. A fluid designed for the specific type of formation and stress conditions of the subsurface environment is injected into the formation at a rate that is greater than the leak-off rate and at a sufficient pressure to overcome the matrix strength and stress conditions within the formation. The rock fractures along a plane that is perpendicular to the weakest, or minimum, compressive stress in the formation. The width of the fractures increases as a function of the injection pressure and physical properties of the formation. When the fracture is wide enough, hard particles (such as graded sand) are added to the frac-fluid injected into the fracture. The particles then migrate into the fractures and hold them open after completion of the well.

As frac-fluid and proppants are injected, the fractures increase in length, height, and depth until the injection is stopped. The height and depth of the fracture within the producing formation are usually stopped by overlying and underlying structures. This confines the fracture to the formation of interest. The length of the fracture away from the wellbore also may be stopped by fall-out and accumulation of proppant at the bottom of the fracture. The growth of the fracture will stop altogether when injection is stopped, or when the leak-off of frac-fluid into the formation becomes equal to the rate of fluid that is being injected. When it is determined that sufficient proppant has been injected to maintain the well at the desired rate of production, the process is stopped. The pressure within the fracture reduces to the original formation stress conditions while the fracture closes on the proppant.

### 1.6.1 Summary of the Shale Hydraulic Fracturing Process

Sedimentary basins have alternating layers, or beds, of shale between unconsolidated sands and cemented sands that form sandstones, carbonate rocks (limestone, dolomite, chalk, etc.), and salts (sodium chloride, gypsum, etc.) that extend to the ultimate depths of the accumulated sediments. Shales are composed of various types of clay containing as much as 50% of other sediments and organic materials mixed with the clays. There are large areal deposits of shales all over the world, including beneath the land-area of several states. The thickness of the shale beds vary from thin lenses of a few inches to several hundred feet. Many of the shale beds contain vast amounts of gas locked in pores that are not connected and thus cannot flow to a low-pressure center, such as a well, for production.

Hydraulic fracturing creates a network of interconnected fractures in the shale that opens channels for the flow of gas from the formation into the wells. The advent of horizontal drilling coupled to hydraulic fracturing has changed unproductive shale gas deposits into large natural gas fields all over the world.

In the United States, the Haynesville and Fayetteville shales are large deposits in Louisiana, Texas, and Arkansas; the Barnett shale in central Texas; the Marcellus shale in the Appalachian basin; the Balken Shale in North Dakota; the Niobrara Shale that extends into multiple states: Colorado, Kansas, Nebraska, and Wyoming; and many others make up the vast gas-shale reservoir.

Vertical wells are first drilled to the shale deposit containing gas (and sometimes oil) and then the drilling is changed from vertical to horizontal and extended, sometimes thousands of feet horizontally, into the shale bed. Multiple horizontal wells may be drilled radially from the initial vertical well and each horizontal spur is then fractured. Consider that a 50-foot hydrocarbon formation will have a producing length of only 50 feet if exposed only to a vertical well, but by extending horizontal wells as much as 3,000–5,000 feet horizontally into the formation, the producing length in contact with the formation becomes 50 times longer. Additionally, fracturing the formation at intervals within the producing zone increases the area exposed for production several thousand times.

After the production tubing (pipe) is in place (in some cases a mile or more in length), sections of the pipe are isolated with plugs and the adjacent formation is fractured through perforations of the pipe in the section. A specifically formulated aqueous solution is pumped into the reservoir under sufficient pressure to produce a network of interconnected fractures extending away from the wellbore and interacting with the natural fractures existing in the shale. Sand or some other granular material (aluminum oxide beads, ceramic beads, sintered bauxite, etc.) is then deposited in the fracture using a modified fluid with a viscosity sufficient to maintain the propping agent in suspension until it is deposited in the fracture where the pressure is diminished. If a proppant is not placed in the fracture then it will close as soon as the frac-fluid is pumped out of the fracture and consequently will not conduct fluids to the wellbore.

After the proppant has been placed in the network of the fracture system, some of the frac-fluid (mixed with formation brine that may be present) flows back out of the well to the surface. This back-flow is either treated for re-use as frac-fluid or injected into a deep brine formation.

The companies generally do not disclose the specific chemicals used to prepare the frac-fluids for competitive reasons, but regulatory bodies have released the principal compositions of the frac-fluids, and a definitive list was published in a Monogram of the Society of Petroleum Engineers (Veatch et al., 1989):

- *Viscosity increase*—natural (organic) thickeners. The organic polymers guar and xanthate gums (some used in food products) can thicken the water solution to the consistency of a gel at ambient conditions.
- *Friction and turbulence reduction of the fluid as it is pumped into the formation*—Several types of polyacrylaminde polymeric compounds have been found to reduce pressure and loss of fluids due to friction and turbulence when pumped through

pipes. These are used in many industries. Friction reduction can promote enormous pumping rates up to 100 bbl/min.

- *Corrosion control and scale inhibitors*—Ethylene-glycol and hydrochloric acid are two chemicals used for this.
- *Biocides*—Methanol and naphthalene are added to control aerobic bacterial that can form metabolic products (such as slimes) and destroy additives used in the frac-fluid in storage and mixing tanks.
- *Surfactants used to enhance proppant carrying ability*—butanol and ethlylene monobutyl ester are two that are used.
- *Control of pH*—various simple bases and acids.

## 1.7 Environmental Issues Related to Hydraulic Fracturing

Frac-fluids consist mostly of water (>95%) with chemicals described above making up less than 2% of the fluid. However, some of these chemicals can be harmful to humans and the environment even in small quantities (parts per billion levels). Therefore, fracture treatments are closely controlled by regulators because of several serious environmental issues. Surface spills of frac-fluids and their chemicals from surface operations, tanks, broken pipes, etc. could allow chemicals to seep into groundwater and surface streams. Well casing may fail (collapse or corrode) and allow pressured fluids to flow upward in the resulting annulus between the pipe and drill-hole, contaminating shallow fresh water aquifers.

Though hydraulic fracturing is commonly conducted at several thousand feet, if it is not carried out carefully it can have a significant negative effect on the environment:

- *Threat to the Underground Source of Drinking Water*—Groundwater, that is, an underground source of drinking water, is generally within a few hundred feet of the surface, but the fracture operation takes place thousands of feet below the surface. Therefore, the possibility of the fracturing fluid contaminating the underground drinking water through a fracture channel is negligible. However, the fracturing fluid can contaminate the groundwater if the integrity of the surface facilities where the fluids are mixed and stored is compromised and there is spillage or leakage on the ground that could percolate into the groundwater or flow into a surface water body (such as a stream) from whence it can be recharged into the groundwater.

Contamination of groundwater can be an irreversible damage to an underground water reservoir. Contamination may also occur if the fractured formation is shallow and there is a fault zone in the vicinity that can act as a conduit for the fracturing fluid. The possibility of this happening is remote due to the natural cementation associated with the faults that hold them in place and restricts any movement of fluid along the fault line. However, if the integrity of the casing and cementation associated with the well are compromised, injected frac-fluid can contaminate the underground water resource.

- *Stress on Groundwater*—Fracturing of each horizontal well generally requires between 3 to 7 million gallons of fresh water that is often taken from an underground aquifer. This high demand on the aquifers can create water shortages (especially in arid regions or regions facing drought conditions) if proper frac-fluid recycling measures are not taken. Withdrawal of huge quantities of groundwater from shallow aquifers also can cause subsidence of the surface area above the aquifer by collapse of the reservoir. Recycling of a portion of the produced water is conducted and a considerable research effort is currently underway to develop technologies for treatment of used frac-fluid so that it can be recycled, decreasing the demand for fresh water.
- *Tremors and Earthquakes*—The possibility of a fracture causing tremors or earthquakes on the surface is extremely remote due to the great depth at which fracturing is conducted. However, flow-back water produced from the fractured formation, which is about 10% to 20% (and sometimes up to 70%) of the injected volume and contains the frac-fluid along with formation brine, is generally injected into injection wells for safe disposal into deep formations. If the injection well is in the vicinity of a fault, it can gradually cause the fault to slip, resulting in tremors and small earthquakes on the surface. Locating injection wells in proper geologic formation is critical in avoiding such incidents.
- *Soil Erosion*—Employing large numbers of heavy trailers and trucks required for fracture treatment can cause substantial soil erosion and damage sensitive flora and fauna of the area. Protection of these sensitive environments can be achieved by careful planning and construction of temporary access roads.
- *Air Pollution*—During the fracking and flow-back process, several pollutants can be released to the air, including methane,

oxides of nitrogen, oxides of carbon, and particulate matter generated from the fractured formation and operation of pumps, generators, and vehicles. Air pollution can be minimized by capturing gas from the formation and using equipment with low emissions.

- *Noise*—The fracturing process requires operation of several high-powered pumps along with mixing of proppants that generates loud noise. Therefore, it can be a source of significant noise pollution, especially if conducted in the close vicinity of population centers. Noise barriers, along with noise mufflers, on the pumps can help reduce some of this.

CHAPTER 2

# Evaluation of Gas-Shale Formations

## 2.1 Sedimentary Formations

Rocks have been placed in three groups based on their origins: (1) *igneous* rocks crystallized from molten magma that are divided between *intrusive* rocks that cool in the subsurface and *extrusive* (volcanic) rocks that cool on the surface; (2) *sedimentary* rocks composed of particles derived from erosion of surface rocks that are transported to a low-lying area where accumulation is possible (such as zones of subsidence, lakes, embayments, ocean shelves, etc.), where the particles undergo burial and subsequent compaction and cementation; and (3) *metamorphic* rocks that develop from mechanical, thermal, or chemical changes of igneous or sedimentary rocks.

Specific types of sediments collect in layers, or strata, and are the end products of differing environmental conditions, such as erosion, transportation, and the settling of particles that make up each different stratigraphic layer. Some examples include the varying layers (with respect to depth) of sandstone, shale, unconsolidated sands, carbonate rocks, etc. The boundaries between the strata are called bedding planes and constitute a thin gradation from one sediment to the other with sharp contrasts between the layers of particle size, composition, compaction, cementation, and rock formation.

The sedimentary rocks can usually be divided into *detrital* and *chemical* (or biochemical) rocks as shown in Table 2–1. Clay is a detritus material formed by the weathering processes that break down rocks in many ways. Mechanical disintegration takes place with the freezing and thawing cycles of water in crevices and pores of rocks, disruption by plant roots, burrowing animals, and chemical changes. Chemical weathering occurs by oxidation of minerals, reaction with

**Table 2–1** General Classification of Sedimentary Rocks

| Detrital | | Chemical/Biochemical | |
|---|---|---|---|
| Type | Diameter | Type | Composition |
| Boulders | 250–2,000, mm | Limestone | Calcite, $CaCO_3$ |
| Cobbles | 60–250 mm | Dolomite | $CaMg(CaCO_3)$ |
| Pebbles | 2–60 mm | Salt (halite) | NaCl |
| Sand/sandstones | 60 μm–2 mm | Gypsum | $CaSO_4{\cdot}2H_2O$ |
| Silt/siltsones | 4–60 μm | Chalk | $CaCO_3$ |
| Clay | < 4 μm | Coal | Carbon |

carbonic acid (formed from carbon dioxide in the air and water), absorption of water, solution in water, etc. The resulting clay is composed of hydrous aluminum silicates interspersed with a minor amount of other minerals. For example, red clay is formed from meteoric dust and fine ferruginous (iron) particles carried by winds or water suspensions. There are three types of feldspars that contribute to the genesis of the various types of clays: (1) *plagioclase* (calcium aluminum silicate $CaAl_2Si_2O_8$), (2) *albite* (sodium aluminum silicate $NaAlSi_3O_8$), and (3) *orthoclase* (potassium aluminum silicate $KAlSi_3O_8$). These react with natural acids in the ground water and soils to form clays. An example is the reaction with carbonic acid:

$$2NaAlSi_3O_8 + H_2CO_3 + H_2O \rightarrow Na_2CO_3 + 4SiO_2 + \underline{Al_2SiO_5(OH)_4}$$

The platelets of clay are flat and less than 1/256 mm in size, thus they are easily suspended in bodies of water and carried to low-lying areas (such as lakes, swamps, tidal flats, and river embankments) where they are deposited along with organic debris that becomes the source of gas and oil when it is rapidly buried in anaerobic (oxygen depleted) zones. The platelets of clay are compacted (as additional sediments accumulate on top) and buried as the depositional area undergoes tectonic subsidence. The clay minerals, mixed with other fine particles and organic debris, solidify into layers of shale as the overburden pressure of overlying sediments increases. The increase of pressure and temperature as the material is buried deeper transforms the organic constituents into a mass called kerogen. As burial continues, the kerogen breaks down into oil and finally cracks and transforms into gas. The average overburden pressure of sedimentary layers increases by about 1.0 psi per foot of depth, and the world-wide temperature gradient is approximately 1°F per 100 feet of depth. The depth of burial in a basin can exceed 30,000 feet, causing large formations of shale containing commercial quantities of hydrocarbons to occur at depths exceeding a mile where the fluid pressure can be 3,000 psi or more and the temperature greater than 150°F.

A prevailing theory of the formation and accumulation of gas and liquid hydrocarbons in subsurface geologic formations is that large amounts of organic material, mixed with finely divided clastic sediments (silt, mud, and sand), is carried by rivers, and deposited in shallow marine environments and river deltas. This material mixed with organic debris derived from aquatic systems (river deltas, swamps, shallow shoals at the edges of large bodies of water) is gradually buried along with the mud, silt, and clay carried in the streams. The amount of fine organic material that accumulates in thick beds (eventually recognized as formations of shale) becomes the source rocks of large volumes of petroleum products that migrate to porous geologic traps. Aerobic bacteria that accompany the sediments as they accumulate destroy the organic matter by using it as a carbon source in their metabolism, but in a quiet swamp-like environment, the dissolved oxygen is quickly depleted, producing an anaerobic environment that preserves the organic material and changes it into kerogen. Absolute anaerobic conditions prevail at a depth of about 50 feet (see Figure 2–1).

Shale is defined as a composite of organic and inorganic sediments composed of material originating from plants, animals, and fine particles eroded from surface rocks and transported to a planar, or bowl-like area, by wind, water, or ice (such as glaciers). We see the deposits on the land surface as quiet aquatic areas, swamps, mud, banks of stream gravel, sand bars, beaches, or sand dunes. When the movement of these particles ends and other particles are accumulated on top, they are gradually buried by the weight. Chemical and physical changes gradually take place as the depth of the sediments increases. Near the surface, where free oxygen is available, chemical oxidation produces changes in the molecular composition of the organic constituents and aerobic bacteria manufacture enzymes (chemical compounds that act as catalysts to breakdown organic molecules). The enzymes break the large organic molecules into smaller components that are then used by the bacteria for growth and energy with ultimate production of water, hydrogen, methane, and carbon dioxide. As the depth of burial increases and the oxygen is depleted, obligate anaerobic bacteria metabolize the residues of organic matter,

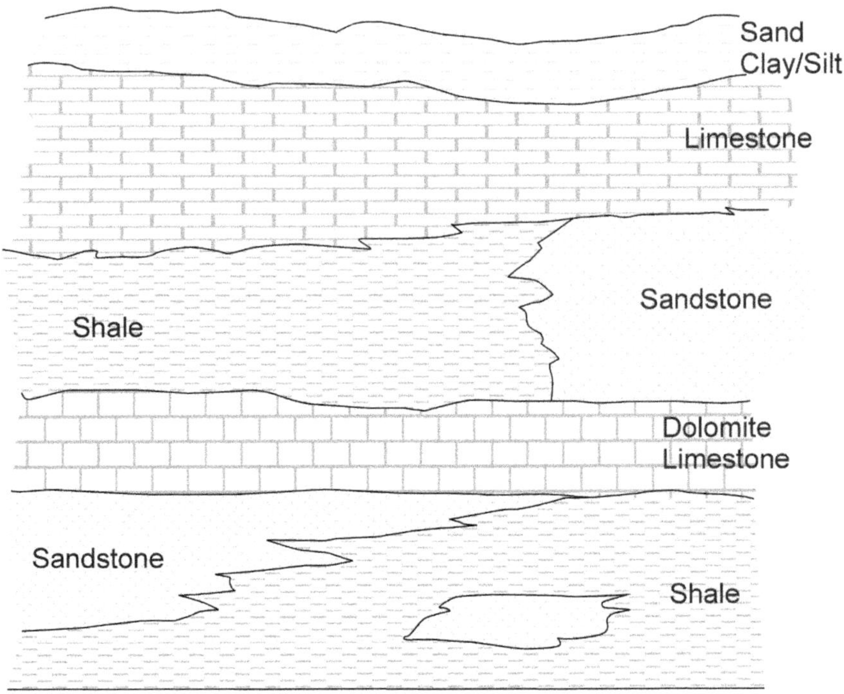


**Figure 2–1** *Burial of successive layers of sediments produced by variations of the surface environment and subsidence.*

converting it into a mass call kerogen, which is a precursor of natural gas and oil. If sulfur is present, anaerobic bacteria will reduce it to hydrogen sulfide, which is blended with methane and other microbial generated gases to give swampy areas the pungent odors recognized as marsh gas.

The inorganic material also undergoes considerable transformation through chemical and physical changes following deposition and burial. The increasing temperature and pressure as the depth of burial increases, combined with water and a host of minerals (that act as catalysts), bring about a wide array of physical and chemical changes. Compaction of the grains induces cementation (and thus rock formation) with the aid of calcium carbonate, silicates, and other minerals. Redistribution of atoms brings about changes of crystalline structures of inorganic matter, and the originally loose sedimentary particles of clay and silt are compacted into hard shales, as well as sands into sandstones.

With respect to gas-shale, its genesis begins with mechanical and chemical degradation of igneous and metamorphic rocks (Tiab and

Donaldson, 2003, 2012). Mud and silt (containing a high load of feldspars, which are the most abundant of all minerals), together with organic matter and sand, are largely dehydrated by compaction, and the mixture undergoes digenesis to form small lens-like structures within other sedimentary rocks and large formations of shale. It is ubiquitous throughout the world and variable in general composition, but easily identified by its physical appearance and by its NScharacteristic response on the spontaneous potential and gamma-ray well-logs. Gas-shale has been classified into four types based on the origin of the organic matter that formed the kerogen, which was later transformed into oil and gas at the depth of burial, increasing the pressure and temperature of the shale: Type I is composed of kerogen that originated from amorphous algae; Type II organic material originated from a mixture of algae and herbal organic matter; Type III developed from woody products and coal; and Type IV gas-shale is characterized as being composed entirely of inert materials. Types I and II transform first into oil, followed by cracking of the oil to light gas, and Type IV changes directly to gas.

The average overburden pressure of about 1.0 psi/ft of depth is composed of the litho-static pressure (grain-to-grain stress of the rock matrix) and hydrostatic fluid pressure (≈ 0.46 psi/ft). The temperature and pressure gradients as depth of burial increases impose "pressure cooker" conditions on the organic matter, gradually converting it from kerogen to oil and gas. The hydrocarbons are expelled from the source rocks into porous rocks, or conduits of porous rocks that are instrumental in promoting the migration of the hydrocarbons to traps where they accumulate, or to the surface where they are found as oil seeps, or tar pits, such as the Brea tar pits in Southern California. When oil enters the vadose zone (where oxygen and aerobic bacteria are present), the lighter components of the oil are consumed and evaporated, leaving behind a viscous tar or heavy oil that continues to deteriorate into a semi-solid material called bitumen.

Depending on the conditions, shale can become the source rock of petroleum products. The oil is eventually squeezed out of the shale beds to migrate into porous sandstone and carbonate rocks. Reserves of gas that are trapped within fine (microscopic) pores were known to exist in large shale beds from the beginning of oil and gas production from sandstones and carbonate formations. However, the compacted, rock-hard shale (as a consequence of compaction of the very fine flat particles of clay) has extremely low matrix permeability, ranging from $10^{-6}$ to $10^{-2}$ md (compared with gas and oil producing rocks with permeabilities greater than 10 md). Therefore, it was not until the development of horizontal drilling technology, coupled with unique fracturing methods and micro-seismic monitoring, that production of shale-gas became a viable commercial product. The gas-shale may have an overall pore volume as great as an oil reservoir in sandstone or carbonate, but the constraint to production of the gas occupying the pores is due to its low permeability. The pores are very small in diameter and restrict the flow of gas.

## 2.2 Shale Formations

The ability of shale bodies to act as a commercial reservoir of natural gas is based on its network of fine pores and fractures. The very small pore sizes in shale result from compaction of the fine grains of silt and clay that are the basis of shale composition with permeabilities that range from $10^{-6}$ to $10^{-2}$ millidarcies (md). Even though the shale may contain a high saturation of gas, the extremely low matrix permeability restricts the movement of the gas, even when very high pressure gradients between the shale matrix and producing well exist, causing the valuable hydrocarbons to be essentially trapped. Natural fractures increase the overall ability of shale to contain and produce gas by providing permeable conduits. Subsurface rocks, especially those that are brittle, such as the hard compacted shale, contain natural fracture systems that result from tectonic stress and motions that all subsurface geologic systems are subjected to. Some of the gas within the beds is adsorbed on residual organic matter and reactive minerals. This gas will not move when subjected to a pressure gradient and thus is not available for production. Until the advent of controlled horizontal drilling and sequenced multi-stage fracture technologies, shale beds were by-passed and considered useful only as upper and lower seals for gas and liquid hydrocarbon reservoirs. Almost all of the conventional gas and oil production comes from permeable sand, consolidated sandstone, and carbonate reservoirs. The hydrocarbons accumulated in these stratigraphic traps migrated from shale source rocks over a long period of time, millions of years in most cases.

Many coal seams also contain copious amounts of gas that are produced through horizontal wells. These tend to be shallow because coal, under considerable overburden pressure, behaves like a plastic material when compressed and completely closes the pore and fractures. The gas-producing coal beds are too shallow to support extensive hydraulic fracturing, thus this production depends on high matrix permeability and a system of natural fractures.

In order for shale-gas to be produced economically, the limitation of low matrix permeability had to be overcome. It was the development of (1) long (>1,000 feet) horizontal well drilling that increases the effective thickness of the formation, with respect to the wellbore (the parameter $h$ in Eq. (2.1)), and (2) hydraulic fracturing effectively increases the producing radius of the wellbore ($r_w$ in Eq. (2.1)). Consider Darcy's equation for fluid flow as applied to gas and illustrated in Figure 2–2 (Tiab and Donaldson 2012, Ch. 7):

$$q = \frac{7.03kh}{\mu_g zT} \cdot \frac{\left(p_e^2 - p_w^2\right)}{\ln\left(\frac{r_e}{r_w}\right)} \tag{2.1}$$

where

$q$ = gas flow rate (scf/D)*$10^9$

$k$ = permeability (md)

$h$ = thickness of the reservoir (ft)

$p_e$ = pressure at the boundary of the drainage area (psi)

$p_w$ = wellbore pressure (psi)

$\mu_g$ = gas viscosity ($cP$)

$z$ = gas deviation factor (Chart, Tiab and Donaldson 2012, see Fig. 2.16)

$T$ = temperature (°F)

$r_e$ = external, or drainage radius

$r_w$ = wellbore radius

**Example 2–1**

Assume that the following data for a well in a shale bed applies. Prepare a table that shows the gas flow rate ($q$) as a function of effective wellbore radius, $r_w$ = 0.5; 1.0; 10; 100; and 1000 feet. Other constant data: $k$ = 0.01; $h$ = 40; $p_e$ = 3,000; $p_w$ = 2,000; $\mu_g$ = 0.0002; $z$ = –.95; $T$ = 140; $r_e$ = 2,500.

The example shows that as the effective wellbore radius is increased (by fracturing), the rate of gas flowing into the well also increases.

| $r_w$ (feet) | $\ln\left(\frac{r_e}{r_w}\right)$ | $q$ (scf)*$10^9$/day |
|---|---|---|
| 0.5 | 8.517 | 0.083 |
| 1.0 | 7.824 | 0.899 |
| 10.0 | 5.521 | 1.270 |
| 100.0 | 3.210 | 2.180 |
| 1000.0 | 0.916 | 7.672 |

## 2.3 Multistage Fractures

The sequenced multi-stage fracturing practiced for long horizontal wells further multiplies the *effective producing radius* ($r_e$) by many times greater than the single fracture illustrated in Ex. 2.1. Hydraulic fracturing also increases the effective permeability of the shale reservoir by opening a large network of distributed small fractures and channels, connected to the natural fracture networks, for flow to the lower pressure region of the wellbore.

The tremendous advantage of horizontal wells over vertical wells can be illustrated simply by considering a producing formation that requires nine vertical wells in an evenly spaced radial pattern extending into the hydrocarbon reservoir from the vertical well to affect economical drainage of the reservoir (see Figure 2–2). In addition, a considerably greater portion of the formation can be contacted for more complete drainage by extending six horizontal wells from a single site placed in the center of the reservoir. Thus horizontal wells yield increased contact area of the producing zone and *sequenced multi-fracturing technology* of each well increases the effective bottom hole well radius. Additionally, there is less environmental impact from drilling operations since horizontal drilling uses only one drill-pad site instead of nine required for a series of vertical wells. This is especially advantageous for production from ultra-low permeability shales.

Computers and mathematical simulation development have transformed the art of hydraulic fracturing into an engineering science such that the well treatments can be carefully designed to con-

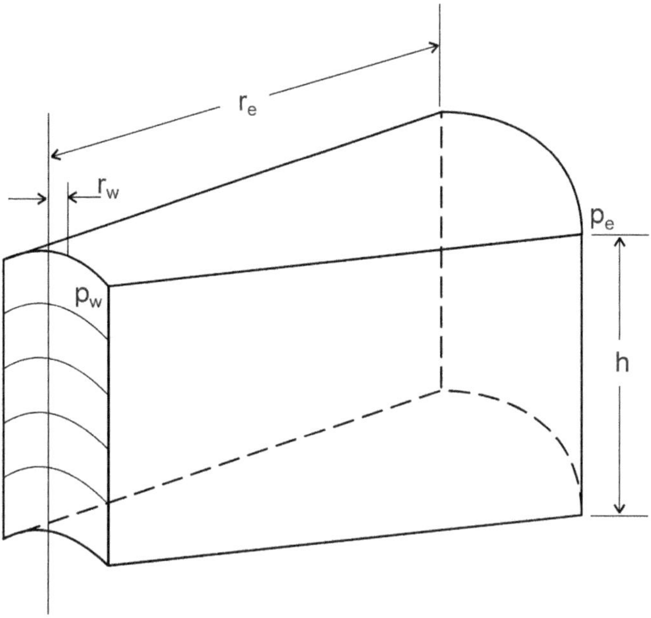


**Figure 2–2** *Parameters used in Darcy's Eq. (2.1) where the subscript e represents the limit of the producing zone where the flow of fluid to the wellbore is zero. The total flow of gas in the wellbore is q (sct/D) and* $r_w$ *is the radius of the well.*

fine the networks of fractures exclusively within the gas-shale reservoirs. Detailed discussion of environmental problems and regulations are reserved for a later chapter.

The fluids used in fracture treatment of shale are generally water-based, containing polymers such as guar gum that thicken the water so that it will more easily suspend and transport the sand or other proppants used to hold the fracture open. Where the formation is relatively deep, "slick-water" is used. Slick-water is just water containing a polyacrylamine polymer that effectively reduces the turbulence of the water being pumped through the conduction pipe, reducing the friction and hence the pumping costs. Shale formations and even zones within a large shale deposit may vary considerably in physical and chemical properties, and therefore specific additives may be added to the frac-fluid to address the differences. Detailed discussion of the major fluids that are currently used is presented later in the book.

After seismic and well-log characterization of the shale formation delineates its boundaries and variations in thickness, a plan for securing cores of the shale bed must be developed and executed to characterize the shale-gas potential. Existing wells that penetrate the shale formation may be used in addition to strategically-located wells designed for retrieval of cores for laboratory analyses.

## 2.4 Fracture Design (Frac-Implementation)

Computers and mathematical simulation yield the expected dimensions (length and height), azimuth orientation, and distribution of fractures that will result from different approaches for achieving the most effective and economical fracture system. Accurate post-fracture evaluation of production rates, with respect to time, yield economic margins that are compiled from accumulation of experience that will be used for future fracture treatments. Analysis of the overall field performance of current wells indicates that the optimum placement of future wells in the formation for maximum production efficiency.

Micro-seismic monitoring and computer simulation are used during the fracture treatment to ensure that the system of fractures does not expand beyond the gas-shale reservoir into porous formations above and below. Such extraneous fractures can act as *thief zones* that absorb the hydraulic fluids and dissipate the pressure that is intended for effective fracture development within the gas-shale reservoir. In addition, extraneous fractures into adjacent non-gas productive formations can funnel great quantities of water into the fractured shale matrix, which will add considerably to costs of gas production caused by additional pumping, separation, and disposal operations. Thus the company will use the best available computer-assisted technology in real time (before, during and after the fracture treatment).

The long (>5,000 feet) horizontal wells that are distributed within the gas-shale formation cannot be fractured through its entire length because the amount of fluid and pumping rates required are too great for that type of operation. Therefore, the fractures are implemented at sequentially staged intervals along the length of the horizontal well using several innovative techniques (or technologies) to accomplish the task with efficiency and minimum costs. Each stage, or location of a fracture treatment along the length of the pipe, is physically isolated and the pipe is perforated. A series of specific volumes of frac-fluids containing fracture propping agents are injected at high rates to open and extend the fractures and to conduct the propping agents into the network of fractures. This is to ensure that the fractures do not close when the operation is completed and the fluid pressure declines to its original value. In shale, the frac-fluids open and extend a host of extant natural micro-fractures, resulting in a complex network of fractures around the selected stage (see Figure 2–4). The number of sub-stage injections, volumes, rates of frac-fluid in each sub-stage, and the amount of proppant to be used are all variables at each stage that are designed to match the specific geological and mechanical properties, formation pressure, temperature, and fluid properties. A single horizontal well may have as many as 10–20 sub-stage injections varying from 10,000–100,000 gallons of injected fluid at rates exceeding 20,000 gallons per minute. The total volume injected may amount to a million gallons of fluid and 500,000 pounds of proppant. Usually the first sub-stage injection consists of a 10–15% hydrochloric acid solution that is designated to clean the wellbore area and perforations that may have been damaged during the cementing and perforation operations. The second injection may be a large volume of slick-water used to initiate the fractures ahead of the more viscous proppant-carrying fluid stages that propagate the network of developing fractures. After the second stage of injection, a number of sequentially decreasing sub-stage injections ranging from 50,000 to 10,000 gallons each are employed. The type of proppant may be changed during the latter stages, and the size of the proppants may be increased.

Thus, the fracture operation is a complex, carefully-planned process that is conducted with computer monitoring throughout the entire process. All of the variables are recorded along with the entire dynamic history of the specific well treatment. Each stage is evaluated in real time in order to incorporate changes for improvement in the sequence of stages such as pumping pressure, rates of injection, and chemical composition of the frac-fluid.

## 2.5 Well Design from Surface to Reservoir

Whether vertical or horizontal wells are to be used, careful attention is exercised to protect shallow zones that contain or transmit low-salinity water and other resources. The standard method uses several sleeves of casing (steel pipe) that are centered in the drill-hole and cemented through the length of the installed pipe (see Figure 2–3). First, a large diameter (16″–18″) pipe is centered in a hole drilled to a depth well below any formation containing low-salinity water than can be used for domestic purposes. Cement is then squeezed into the pipe until it rises upward in the annulus between the casing and the wall of the wellbore all the way to the surface. The cement, bonded to the well casing, effectively isolates the well from the fresh water aquifers. After removal of the cement inside the surface casing, a smaller diameter bit is used to drill to the formation to be used for waste

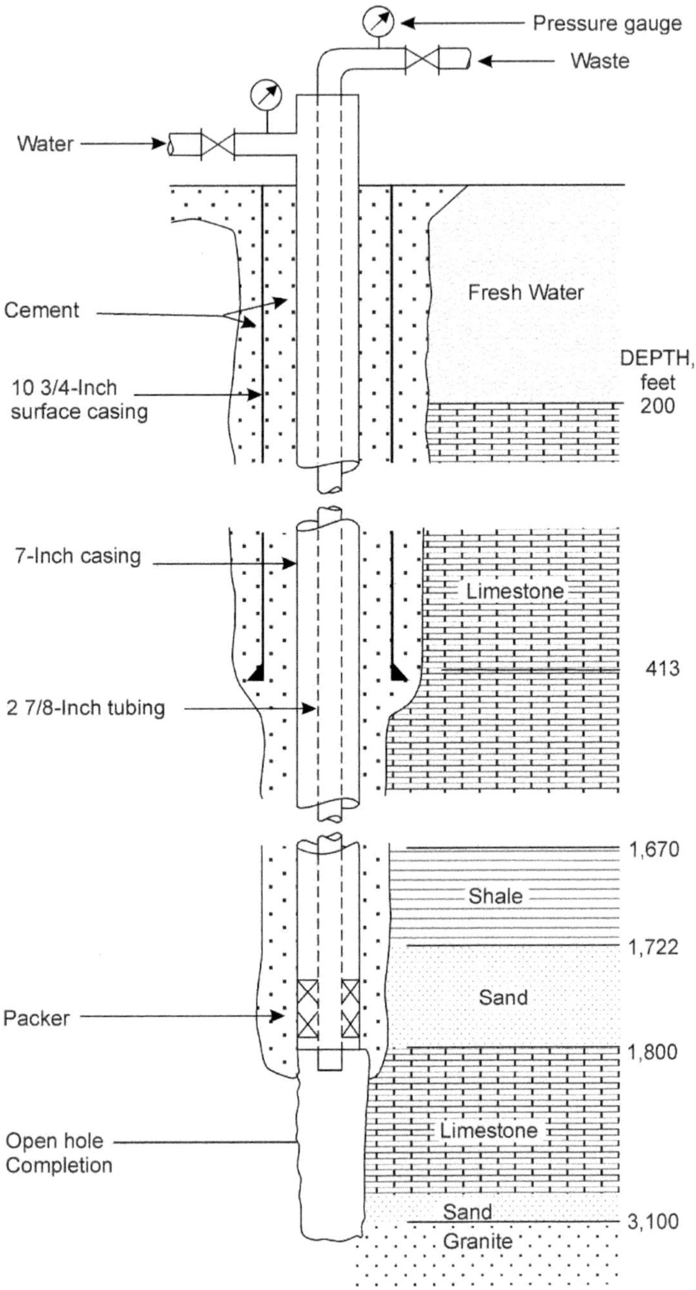


**Figure 2–3** *Design of a typical open hole disposal well for protection of useful aquifers. Water pressure in the annulus monitored for leaks.*

injection or production of hydrocarbons. Casing is inserted and centered then the annulus of the second casing is cemented across zones that contain brine under pressure that would channel downward to the injection (or producing) formation through the annulus and interfere with injection of waste, or production of gas and oil. In deep wells it is not possible (or necessary) to cement the entire length of the wellbore because the pressure exerted by the column of cement in the wellbore would fracture the porous zone and fill the fractures with cement, ruining the possibility of using it for injection of waste or production of hydrocarbons. A liquid waste injection well, however, will be cemented in place through their length, if at all possible, as shown in Figure 2–4.

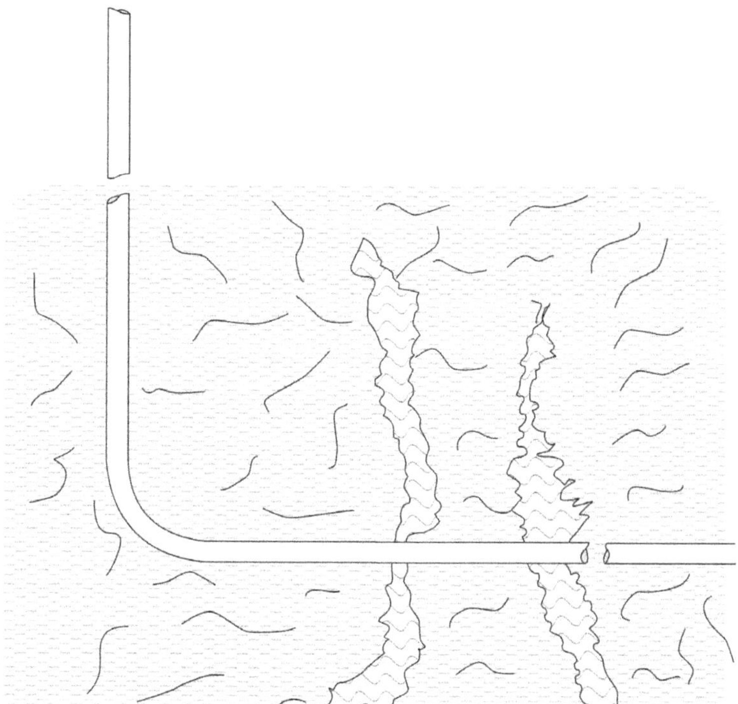

**Figure 2–4** *Two stages of a ball-drop/sleeve fracture system of a horizontal well in a shale bed. The wavy lines w/in the shale indicate micro-fractures.*

### 2.5.1 Back-Flow of Frac-Fluid

When the fracture treatment has been completed and the fluid pressure within the reservoir is relaxed, the pressure in the reservoir exceeds the pressure of the column of water filling the well. The column of fluid filling the pipe exerts a pressure gradient that is a function of its density and is equal to around 0.5 psi/ft of depth. Thus for a 5,000-ft well, the pressure at the bottom would be approximately 2,500 psi. If this pressure is less than the residual pressure left in the formation at the end of the fracture treatment, fluid in the reservoir will move to the surface as *back-flow.* Anywhere from 25–75% of the injected frac-fluid will flow back to the surface and it will be mixed with variable amounts of formation brine. The subsurface formation water may contain salts (generally sodium chloride, calcium chloride, and smaller concentrations of many other salts) ranging as *total dissolved salts* (TDS) from fresh water with less than 5,000 ppm (5%) TDS to salt-saturated brines containing 200,000 ppm (20%), or more, TDS. The concentration of salts in formation brines will certainly vary greatly from one shale deposit to another, and it also may vary in different areas of the same formation.

Frac-fluid back-flow also is stimulated by adding high pressure carbon dioxide or nitrogen to the frac-fluid. This is especially necessary where the reservoir pressure is not high enough to expel the frac-fluid when the fracture treatments have been completed. The entrained gas causes rapid removal of the spent frac-fluid and has the added advantage of diminishing frac-fluid loss to the formation matrix by displacement of reservoir fluids.

The selection of nitrogen or carbon dioxide for the energizing gas is based on the quite different physical behavior of the gases. Nitrogen is insoluble in the frac-fluids and a portion of the nitrogen tends to leak-off into the formation, but as soon as the treatment is completed and the frac-fluid pressure relaxed, the remaining nitrogen evolves rapidly from the fluid in the fracture network and immediately generates pressure as the gas expands to cause flow-back of the spent frac-fluid.

In contrast to nitrogen, carbon dioxide is quite soluble in the frac-fluid mixture and also forms carbonic acid when in contact with water. The dissolved $CO_2$ does not migrate into the formation, so there is no loss due to leak-off that is characteristic of $N_2$. The energy is stored in frac-fluid by $CO_2$ and when the pressure is reduced at the end of the treatment, a solution gas-drive mechanism develops, effectively removing the frac-fluids from the fractures and enhancing the flow-back.

The large volume of back-flow fluid occurs at diminishing rates over a period of days, or weeks, and continues as the well is placed on gas production. The highly variable mixture of back-flow fluid must be captured and disposed in such as way that it does not result in spills on the surface where it would cause a severe pollution problem.

There are a number of options available for treatment and disposal of back-flow fluids. The procedure that is adopted depends on the economics and local conditions. If a large gas-shale formation has been discovered, portable equipment may be used in the field for on-site treatment and re-use of the fluid after blending and dilution with new make-up water. The chemical composition of the back-flow, amount, and volume determine the treatment required to produce the quantity of water that can be reused. Adjustment of pH to neutral followed by filtration for removal of entrained and precipitated solids may reduce the undesirable chemicals to a concentration that is insignificant (for frack-treatment) after the water is blended with fresh make-up water. Deliberate precipitation and removal of solids may be possible by addition of a base such as sodium aluminate or sodium hydroxide followed by flocculation and filtration. Dilution with make-up water will reduce the cost of subsequent pH adjustment. With facilities on site for such treatment, cost effective chemical additives, recommended by the laboratory, can be used for total dissolved solids removal and neutralization to reclaim the returning back-flow water.

If the back-flow is contaminated with a high concentration of salts from the formation water making chemical treatment and dilution impractical, the fluid can be treated by distillation on site. Distillation requires a considerable amount of heat energy and specially-designed units for field use, but in some circumstances it can be the most economical option. Distillation, however, has the advantage in that it produces completely purified water that is available for any use whatsoever. The by-product salts are a small amount that is sometimes useful to another industry or that can be easily transported for safe disposal.

In cases where reuse of the back-flow fluid is practical or economical, chemically treated and neutralized water may be used to relieve agricultural areas impacted by drought. If the gas field is near a large municipal community, arrangements may be made for construction of a temporary pipeline to transfer the back-flow itself, or pretreated back-flow, to the municipal treatment/disposal facility, or a commercial water treatment facility may be in reach that will process the back-flow for a fee based on its chemical composition and volume.

Finally, the back-flow may be treatable in the field to a purity that makes it permissible to be discharged to a surface stream.

If gas-shale is within, or near to, an oilfield where injection wells are allowed by state and federal regulations for disposal of the oilfield produced brines, the back-flow may be transferred to the oilfield disposal facility and blended into the produced brine destined for subsurface injection into a brine aquifer. The petroleum industry has a long history of treatment and reinjection of oil reservoir brines that are inadvertently produced along with oil. The US Bureau of Mines investigated the use of injection wells in the 1960s for disposal of industrial chemical liquid wastes. This thorough investigation resulted in an upsurge of the chemical industry's use of injection wells and establishment of proper design criteria as well as subsequent federal regulations instituted by the Environmental Protection Agency (Donaldson, 1964; Bayazeed and Donaldson, 1970, 1971; Donaldson, 1972; Donaldson et. al., 1974).

## 2.6 Gas-Shale Reservoir Characterization

Shale generally contains natural fractures, some of which are connected throughout the complex networks of fractures. This complexity interferes with standard well-log interpretations and therefore the well-logs are dependent on calibration by laboratory analyses of the petrophysical parameters. The large amount of petrophysical properties that are required for determination of the shale-gas potential for economic production, horizontal well drilling, and production stimulation by hydraulic fracturing depend on the core sample and well-log analyses of the potential gas-shale discovery. These analyses are too complex and inter-related to be conducted without computer programs that can simultaneously evaluate the information recorded on the various logs and integrate the log-derived data with laboratory analyses coupled to statistical interpretations. These programs and simulation models are beyond the scope of this book. Basic well-log interpretation and programs are presented by Donaldson and Tiab, 2012. Explanation of the petrophysical data that are required and their use are discussed.

Classical well-log interpretation cannot be applied directly to shale formations because of the wide variability of the mineralogy, fine pores and natural fractures, organic composition, and overall electrical conductivity. These variables present complexities that vary in zones throughout the gas-shale structure. Therefore, the interpretation of well-logs depends on laboratory petrophysical analyses for calibration; this dependence has been labeled *ground truthing laboratory analysis*.

As stated earlier, shale is composed of small grains of clay minerals, (30% or more) such as smectite, illite, and kaolinite. The balance of the materials are clay-sized (≤1/256 mm) mineral particles of feldspars, quartz, chert, pyrite ($FeS_2$) and other sulfide minerals, iron oxides, carbonates, and many other minerals that are functions of the depositional environment. As the deposits accumulate, they are mixed with varying amounts of organic materials that are later accumulated aq an organic mass called kerogen as the sediments undergo burial. The increasing temperature and pressure as burial continues, aided by catalytic behavior of some of the minerals, transform the kerogen into hydrocarbons. In the case of gas-shale, the kerogen was modified into light natural gas (methane, ethane, propane, butane, and traces of pentane). The gas occupying the pores of shale is practically immobile because the pores are very fine and consequently the permeability is a limiting small value (<0.01 md in most cases). Thus, the only way to impart mobility to the gas is to develop a network of fractures that interact with the micro and macro natural fractures to impart conduits for movement of the free and adsorbed gas within the matrix when a lower pressure than the normal reservoir pressure is established within the well. Kerogen is part of the total organic content of the gas-shale and consequently decreases the total electric conductivity of the formation. This, in turn, can present a false assumption of much greater gas saturation than the amount of gas actually present. Consequently, cores from the zone within the shale that is under consideration must be analyzed for determination of a statistical mean amount of kerogen present; even this analysis is complicated by the presence of absorbed gas within the kerogen. Laboratory procedures, however, allow determination of both adsorbed and absorbed gas.

In 1942, Archie published a paper relating the water saturation ($S_w$) of a subsurface reservoir to porosity ($\varphi$), the resistivity of the water in the pores ($R_w$), and the total resistivity of the rock ($R_t$). Archie's equation immediately transformed resistivity well-logs from a qualitative indication to the presence of gas and/or oil in sandstone and carbonate formations to a quantitative determination of the hydrocarbons in place. The effect of the electrical conductive of the clays present in the shale was corrected by subtraction of the proportional deflection produced by the conductivity of the shales ($V_{sh}$, or volume percent of shale; $V_{clay}$ will be used instead to avoid confusion) from Archie's equation:

$$S_w^n = \frac{aR_w}{\phi^m R_t} - V_{clay} = F_r \frac{R_w}{R_t} - V_{clay} \tag{2.2}$$

The negative charges of clays bond principally with the cations in the brine (calcium, potassium, sodium strontium, etc). However, in cases where the formation water has low salinity (< 10,000 ppm), it is the polar water molecule that is bonded to the negative charges on the clay. This forms a cation layer with increased volume and electrical conductivity, which in turn yields a false lower resistivity of the formation, indicating that there is a lower saturation of hydrocarbons than is actually present (Donaldson and Alam, 2008). For gas-shale it is imperative to measure the actual water saturations of cores from the zone of interest in order to properly calibrate the resistivity well-log. This calibration is especially necessary where the water saturation is very low.

Greater accuracy for interpretation of well-logs of sandstone and carbonate reservoirs containing gas and oil became possible with the introduction of the Simandeaux and Waxman-Smit equations that incorporate terms for subtraction of the detrimental influence of electrical conductivity of clay minerals. The Simandeaux equation is applicable to a wide range of reservoir conditions and is simply an expression of Archie's equation as a quadratic equation:

$$\frac{1}{R_t} = \left(\frac{V_{clay}}{R_{sh}}\right) S_w + \left(\frac{\phi^m}{aR_w}\right) S_w^2 \tag{2.3}$$

The Waxman-Smit equation introduces a correction of the value of the formation resistivity factor using the ion exchange properties of the clays present in the shale. The Waxman-Smit equation is considered to be the most accurate and is used when data on the ion exchange properties of the clays are available from laboratory measurement. Both equations and examples of their use are presented by Tiab and Donaldson, 2012. The corrected formation resistivity factor is:

$$F^* = \left(\frac{1}{C_o}\right)\left(C_{eq} + C_w\right) \tag{2.4}$$

Generally, Archie's equation is not affected by high salinity, natural fractures, or low values of permeability.

Capillary pressure measurements of the pore throat size distribution show that the pore throats of gas-shales are less than 1µm. The water content is immobile due to the small pores that have very high capillary pressure physically retaining the water. This capillary-bound water introduces a path for electrical conductivity that lowers the true resistivity of the reservoir matrix, which in turn translates to false indication of higher water saturation. Hence, the nuclear magnetic resonance (NMR) log is used to determine the amount of capillary pressure-bound water and can be used directly as input to the computer logging program (for real-time analysis) to correct the resistivity log water saturation.

The major suite of logs that is run for gas exploration in shale beds is the triple combination: gamma-ray, resistivity, and density/neutron logs.

## 2.7 Gamma-Ray Well-Log

The gamma-ray well-log detects gamma rays emitted by the radioactive elements potassium, thorium, and uranium, which are common to all accumulations of clays. Shales contain variable amount of clays ranging from 30 to 60% and therefore emit a greater number of gamma rays than sandstone and carbonate formations. Therefore, the increased, uniform intensity of gamma rays is the unique signature of the shale and in addition offers a statistically valid measure of the amount of clay within the formation. Furthermore, the gamma-ray log can be recorded in open holes or cased wells with any type of fluid in the well. The diminished amount of gamma rays that occurs at each casing collar, where there is a greater density of steel, is used for accurate measurement of the depth of cased wells. The collar locator offers accurate placement of casing perforations, placement of fractures, and other down-hole well completions and remediation.

The gamma-ray log is calibrated in American Petroleum Institute (API) units, which are equivalent to 0.07 µg of radium per ton of rock; average mid-continent shale records about 100 API units. Layers of shale and variable amounts of shale in sands are distinguished by the intensity of the gamma rays recorded, which is used to obtain the gamma-ray shale index ($I_{GR}$) that is used to calculate the quantity of clay (and thus the shale) within the formation. These data must be incorporated in Archie's equation for accurate determination of the water saturation.

$$I_{GR} = \frac{GR_z - GR_{cs}}{GR_{sh} - GR_{cs}} \tag{2.5}$$

The radiation recorded in unconsolidated sands requires computation by a different equation than the one used for consolidated formations because of the difference in transmission of the rays through the rock formations. Hence Eq. (2.5a) is used for friable, or unconsolidated, zones, and Eq. (2.5b) is specific to consolidated clastic-type zones:

$$V_{clay} = 0.083\left[2^{(3.7*I_{GR})} - 1.0\right] \tag{2.5a}$$

$$V_{clay} = 0.330\left[2^{(2.0*I_{GR})} - 1.0\right] \tag{2.5b}$$

## 2.8 Density/Neutron Log

The density log is based on scattering of gamma rays emitted by a gamma-ray source that is focused into the formation. Some of the gamma rays are absorbed by the matrix, and the intensity of those arriving at two fixed distances from the source are recorded. The bulk density of the formation is determined by correlation to a standard concrete containing an amount of radioactive compounds that yield a radiation intensity of 100 API units. Recorded formation density is the average value of the matrix plus the fluids contained in the pores and is recorded on the log directly as g/cm$^3$. The density log is used to determine the porosity of the formation by using the matrix bulk density ($\rho_{b,}$ obtained from cores). The porosity determined by the density log must be compensated for the natural gamma rays emitted by minerals associated with the clays in the formation of interest ($V_{clay}$). Thus the final computation is obtained from:

$$\phi_d = \left(\frac{\rho_m - \rho_z}{\rho_m - \rho_f}\right) - \left(\frac{\rho_m - \rho_{sh}}{\rho_m - \rho_f}\right) \tag{2.6}$$

The neutron log is based on principles of nuclear radiation that are beyond the scope of this book. A neutron source, such as beryllium/radium composite on a wire-line tool, emits a continuous flux of neutrons, and some of the neutrons are absorbed by atoms of the fluid in an open hole, and casing if present, and by the formation. A detector placed about 18 inches away from the source on the same tool measures the amount of radiation arriving. Neutrons entering the formation are absorbed by hydrogen atoms of the pore fluids (water, hydrocarbons, and kerogen), and the measurement at the detector is indicative of the amount of hydrogen in the formation. High porosity results in absorption of a greater amount of neutrons than with low porosity. Thus the amount of neutron absorption by the formation is correlated to the porosity of the formation contacted by the neutrons. The clays in the formation cause distortion of the true values of the porosity because of the natural radioactivity from some elements in the clays; therefore, a correction for the amount of clay in the formation (from the gamma-ray log) is added to the computation of porosity. It is added because the measured porosity is lower when radioactive clays are present. Therefore the neutron porosity is:

$$\phi_n = \phi_z + V_{clay}\phi_{clay} \tag{2.7}$$

The true porosity is determined from the values of porosity determined by the density and neutron logs as follows:

$$\phi_{true} = \sqrt{\frac{(\phi_d^2 - \phi_n^2)}{2.0}} \tag{2.8}$$

## 2.9 Use of Seismic Data

The process of fracturing (or breaking) rocks in the subsurface produces vibrations, and thus acoustic waves, that are detected by geophones and recorded by a seismometer. Drilling through a rock is a noisy event because the vibrations of the down-hole drilling motor (or pipe stem rotating in the hole) and movement of the drilling fluid containing rock fragments are added to the cracking of the rocks. An array of geophones set a various depths can be used to locate the exact position of the drill-bit and follow it as it traverses the reservoir using the GPS coordinate grid and an array of geophones set at various depths at three locations around the well. Likewise, after the well has been drilled and cased, and the casing perforated at locations that promise the best productivity, precise hydraulic fracturing can be conducted using the sounds that emanate as the rocks fracture and turbulence of the frac-fluid within the casing. When pumping stops, noise from the dynamic fracture treatment process ends, but the fracture growth continues at the ends of the new fractures producing micro-seismic vibrations with frequencies ranging from 10 to 10,00 Hertz (vibrations per second). Arrays of three sets of geophones at different depths locate the micro-fractures, which are imaged on a computer in real time to precisely establish the 3-D size, growth, and location of the fractures.

All atoms and molecules contain inherent forces that produce random vibrations, and a force of attraction that holds them together. These forces resist deformation, and therefore they compact under pressure and extend when subjected to a tensile force, within limits. The ability to decrease and increase in volume is a property known as *elasticity*, which is the ratio of stress (applied force per unit area) to strain (degree of deformation with respect to the normal size at rest). There are three types of deformation: length, volume, and angular, which are expressed as the moduli of elasticity, or ratios of stress to strain:

$$\text{Young's modulus } E = \frac{F / A}{\Delta L / L} \text{ (change of length)} \tag{2.9}$$

$$\text{Bulk modulus } B = \frac{F / A}{\Delta V / V} \text{ (change of volume)} \tag{2.10}$$

$$\text{Shear modulus } S = \frac{F / A}{\tan s} \text{ (change of angular shape)} \tag{2.11}$$

The distances between molecules collected in a group differ with respect to each material at any specific temperature and pressure; thus, gases have the greatest separation between molecules, followed by liquids, and solids with the least separation. The compressibility decreases from gases to solids (note that the modulus of elasticity is the inverse of compressibility). When an elastic body is temporarily displaced and released, the action causes the molecules to oscillate in unison at a frequency that is characteristic of the medium (or material) disturbed, and the oscillation will be transmitted spherically from the point of origin as elastic waves that travel at specific velocities. In a solid, the molecules set in motion do not move away from their position; they only vibrate elastically within their central position and generate two principal waves that move through the material. One wave oscillates in a longitudinal mode parallel to the direction in which the wave is moving away from the point of origin; this is a *compression*, or *pressure*, wave. The second principal wave is a displacement that moves up and down (in a transverse mode) to the direction of propagation away from the point of origin; this is a *transverse*, or *shear*, wave. Only solids propagate both compression (*P*-waves) and shear (*S*-waves) waves as well as have the rigidity that allows transverse motion. Gases and liquids cannot transmit waves composed of shear motions.

The velocity of the *P*-wave is a function of both the Bulk modulus of elasticity (*B*) and the sear modulus (*S*); and the density. The *S*-wave is a function only of the shear modulus and density, as follows:

$$u_c = \left( \frac{B + 1.33S}{\rho} \right)^{0.5} \tag{2.12}$$

$$u_S = \left( \frac{S}{\rho} \right)^{0.5} \tag{2.13}$$

An array of geophones on the surface, and at various depths in surrounding wells, channels the micro-seismic vibrations of hydraulic fracture treatments to a central computer. The large amount of data are digitally processed into a three-dimensional image of the exact location and growth of the fractures as they occur. Thus the fracture treatment is minutely controlled and contained within the most productive zones of the gas-shale reservoir. This precision absolutely avoids the growth of fracture into adjacent reservoirs above, laterally, or below the gas-shale zone of interest. Saldungaray and Palisch; and Coates (2012) discuss cases histories of recent applications of the evolving hydraulic fracture technology in gas-shale reservoirs.

The distance to any seismic event is determined by the separation of the *P* (pressure) and *S* (shear) waves that are propagated simultaneously at the site of the event, but the *P*-wave travels at a considerably higher velocity (about 4 miles per second in limestone) than the *S*-wave (about 2 miles per second in limestone). The difference in arrival times of the two waves at any distance from the seismic event is used to calculate the radial distance between the event and the geophone's location, assuming that the average velocities of the waves passing through the intervening soil and rocks are known, as follows:

$$\Delta r = \Delta t * u_c \left( \frac{u_s}{u_c - u_s} \right) \tag{2.14}$$

where

$r$ = radial distance from the center of the disturbance

$t$ = time, seconds

$u_c$ = velocity of the P-wave (compression wave)

$u_s$ = velocity of the shear wave

All of these analyses lead to the design of the fracture treatment program for horizontal drilling. Fracture stage locations are placed at about 250-ft intervals along the length of the horizontally drilled well, and the fluid design and pumping schedules are generally 2,000 to 3,000 gal/ft of shale reservoir thickness with a proppant load about 1,000 lbs/ft of shale at fluid injection rates exceeding 50 bbl/min.

CHAPTER 3

# Rock Mechanics of Fracturing

## 3.1 Introduction

All positions within the subsurface environment are under vertical and horizontal stresses that result from tectonic and overburden forces. Discussion of the host of theories of fracture technology is beyond the scope of this book. Instead, the basic understanding of hydraulic fracturing will be presented and illustrated by examination of practical methods that yield a clear analysis of the art. Two important assumptions will be used: (1) that the subsurface formation is isotropic and somewhat elastic (the linear stress-strain relations of a rock are expressed by Young's modulus and Poisson's ratio).

This means that the pores of the rock can expand and contract to some small degree in response to variation of the pore pressure as expressed by the poro-elastic theories of Geerstma (1953); and (2) that the fluid producing a fracture is an incompressible fluid, and the drilling "mud" filling the drill-hole has a pressure equal to the hydraulic gradient 0.465 psi/ft.

In order to understand the art of hydraulic fracturing, it is essential that the basic technical terms of rock mechanics be thoroughly understood.

## 3.2 Young's Modulus of Elasticity (*E*)

An elastic body will deform when its motion in space is constrained while a force is applied to it. A compressive force tends to flatten, or twist, the body while a tensile force causes stretching, or elongation. Stress is the application of a force over an area of the body and its intensity is expressed as force applied per unit area of the body:

pounds-force per square inch (lbs/in$^2$) or Newtons per square meter ($N$/m$^2$). Strain is the measure of deformation of the body, with respect to an original length or width, which is caused by the applied stress: (a) the change of length with respect to the original length of the body ($\Delta L/Lo$), (b) the change of radius, or diameter, of a cylinder with respect to the original radius, or diameter, of the body ($\Delta r/r_o$ or $\Delta d/d_o$), or (3) change of volume with respect to the original volume ($\Delta V/V_o$), (see Figure 3–1):

$$Young's\ Modulus = \frac{Stress(\sigma)}{Strain(\varepsilon)} = \frac{F / A}{\Delta r / r_0} \quad \left[\frac{lbs / in^2}{in / in}\right]) \tag{3.1}$$

$$Hooke's\ Law \quad \frac{F}{A}(\sigma) = (E)\frac{\Delta r}{r_o}(\varepsilon)$$

Eq. (3.1) may be rearranged to show that the stress of an elastic body is directly proportional to the applied strain where the proportionality constant is Young's modulus; this relationship is known as Hooke's law. If the cylinder shown in Figure 3–1 is an isotropic elastic body, then the lateral strain (deformation of the radius) is proportional to the applied compressive stress (force per unit area) and the slope of the line is equal to Young's modulus (see Figure 3–2).

Rocks are heterogeneous composites of materials containing crystals, cementing compounds, discontinuities, and micro-cracks randomly distributed throughout. Subsurface rocks are subjected to a compressive overburden stress that closes the natural micro-cracks. Consequently, Hooke's law does not apply exactly to rocks; a stress-strain diagram of a rock is S-shaped, containing three distinctly identifiable regions ending at a point where the rock fails (fractures or shatters) as a result of the applied stress (see Figure 3–3). The curvature of Region I results from closure of natural micro-fractures, and as the stress is increased, deformation of the grains yields an almost linear relationship between the applied stress and resulting strain, resulting in Region II. In Region III, new micro-fractures are formed as stress continues to be applied until failure occurs at the end of the curve.

If a vertical force (acting in such a way as to pull the body apart) is applied, it is a tensile force that would cause the body to elongate while the sides contract in response to the tensile force. By convention, tensile force is negative and compressive force is positive.

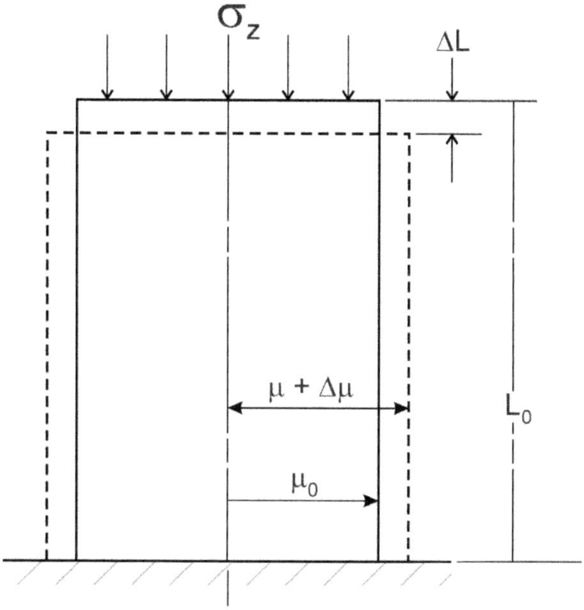


**Figure 3–1** *An elastic cylinder, with no confining lateral stress, resting on a flat surface and subjected to a vertical compressive stress (lbs./in²).*

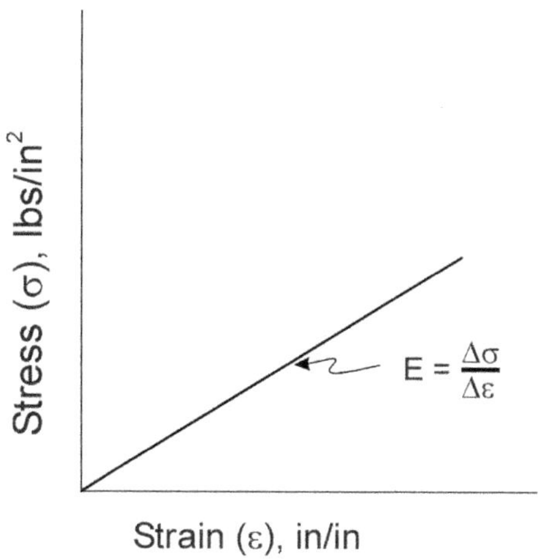


**Figure 3–2** *Hooke's Law. Stress is proportional to strain. The slope of the line is equal to Young's Modulus of elasticity.*

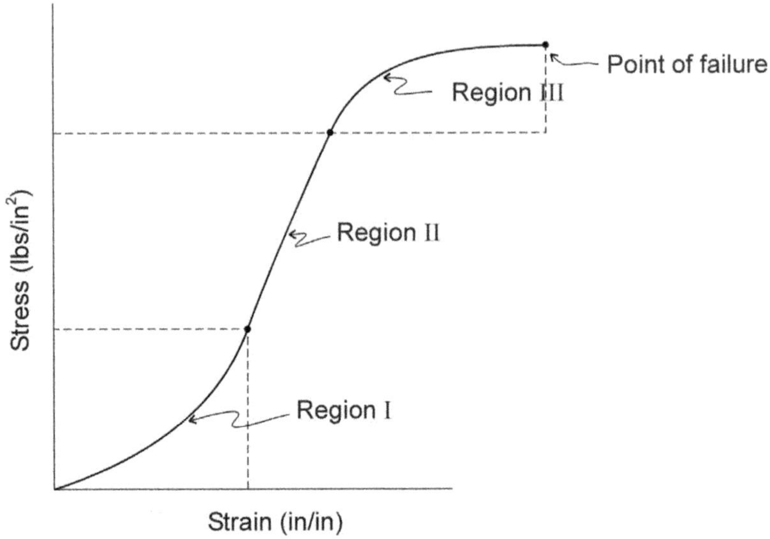


**Figure 3–3** *Stress-strain relationship of a rock. Region I: Plastic strain caused by closure of micro-fractures. Region II: Elastic compression of the rock matrix material. Region III: Plastic strain caused by micro-fracture formation in response to the applied stress until failure occurs.*

## 3.3 Poisson's Ratio ($\nu$)

As shown in Figure 3–1, the applied stress ($\sigma_z$) causes the cylindrical body to expand radially along the radius of a cylindrical core (lateral strain) and diminish in length along the axial length (axial strain). The ratio of the lateral to axial strain is *Poisson's ratio*:

$$\nu = -\frac{\varepsilon_{lateral}}{\varepsilon_{axial}} = -\frac{\Delta d / d_o}{\Delta L / L_o} \left[\frac{in/in}{in/in}\right] \tag{3.2}$$

## 3.4 Bulk Modulus ($K_B$)

The bulk modulus results from a reduction of the *bulk volume* of a body that occurs when equal forces are applied to all sides. This will occur if a water saturated core is placed in a container with rigid walls filled with a fluid. When the exterior fluid pressure surrounding the core is increased, pressure is exerted uniformly on all parts of the body, and thus the entire bulk volume of the body is reduced. The opposite is the expansion of a porous body, which is saturated with a fluid;

expansion of the body will occur when the fluid pressure in the pores is increased. The bulk modulus is thus defined as:

$$K_B = \frac{F / A}{\Delta V / V_o} \left[\frac{lbs / in^2}{in^3 / in^3}\right] = \frac{E}{3(1-2\nu)} \tag{3.3}$$

Bulk modulus is the reciprocal of the compressibility of a body:

$$C_b = -\frac{1}{V_b}\left(\frac{\delta V}{\delta P}\right) = \frac{1}{K_B} \tag{3.4}$$

## 3.5 Shear Modulus (G)

If parallel planes of a body are subjected to forces that move them in opposing angular directions from each other, the energy exerted on the body tries to twist, or tear, it apart. The ratio of the shear stress to the angle formed by the deformation from its original position is the *shear modulus*, which is a measure of rigidity, or resistance of the body to change of its shape (see Figure 3–4):

$$G = \frac{shear\ stress}{Angle\ of\ deformation} = \frac{F / A}{\theta} \left[\frac{lbs / in^2}{radians}\right] = \frac{E}{2(1+\nu)} \tag{3.5}$$

$$radian = 57.3\ degrees$$

The elastic properties of subsurface formations are measured with a long-spaced sonic log (which determines the values of the compression-wave and shear-wave velocities). The log is combined with a density log

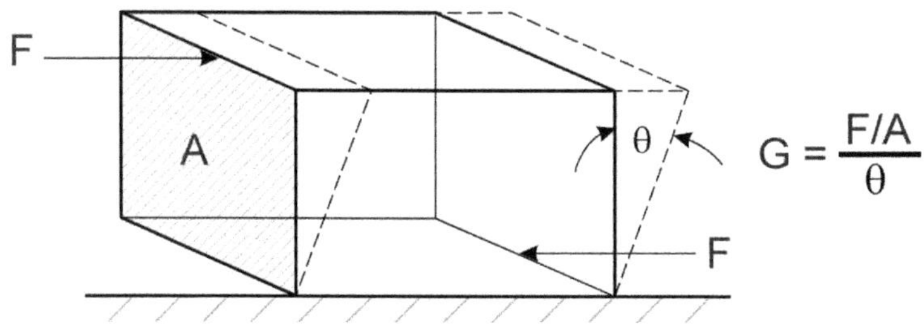


**Figure 3–4** *The shear modulus is the ratio of the shear stress to the angle of deformation (θ, expressed in radians).*

to simultaneously obtain the bulk density of the formation. The elastic properties are then calculated as follows:

$$E = \frac{\rho_b u_s^2 (3u_c^2 - 4u_s^2)}{u_c^2 - u_s^2}$$
$$G = \rho_b u_s^2 \tag{3.6}$$
$$\nu = \frac{u_c^2 - 2u_s^2}{2(u_c^2 - u_x^2)}$$

where

$E$ = Young's modulus

$G$ = shear modulus

$\nu$ = Poisson's ratio

$u$ = velocity

$\rho$ = density

**Example 3–1** Sample Calculation

A uniaxial test (no confining pressure) is conducted using a cylindrical core-holder (see Figure 3–5). Assume a core length/diameter of 2.5 in/1 in, and a vertical load ($\sigma_z$) of 18,000 lbs applied over the surface area of the top of the core. The change in length was $3*10^{-3}$ inches and change of diameter was $5*10^{-4}$ inches measured with strain gauges attached to the core.

$$\sigma_z = \frac{F}{A} = \frac{18{,}000}{\pi(1.0/2)^2} = 22{,}918\ psi$$

$$\varepsilon_z = \frac{\Delta L}{L_o} = \frac{3*10^{-3}}{2.5} = 1.2*10^{-3}$$

$$Young's\ Modulus\ (E) = \frac{22{,}918}{1.2*10^{-3}} = 19.1*10^6$$

$$Poisson's\ ratio\ (\nu) = \frac{\Delta d / d_o}{\Delta L / L_o} = \frac{5 * 10^{-4}}{1.2 * 10^{-3}} = 0.42$$

$$Bulk\ Modulus\ (K_B) = \frac{E}{3(1-2\nu)} = \frac{19.1 * 10^6}{0.54} = 35.4 * 10^6$$

$$Bulk\ Compressibility\ (C_b) = \frac{1}{K} = \frac{1}{35.4 * 10^6} = 0.028 * 10^{-6}$$

$$Shear\ Modulus\ (G) = \frac{E}{2(1+\nu)} = \frac{19.1 * 10^6}{2.84} = 6.7 * 10^6$$

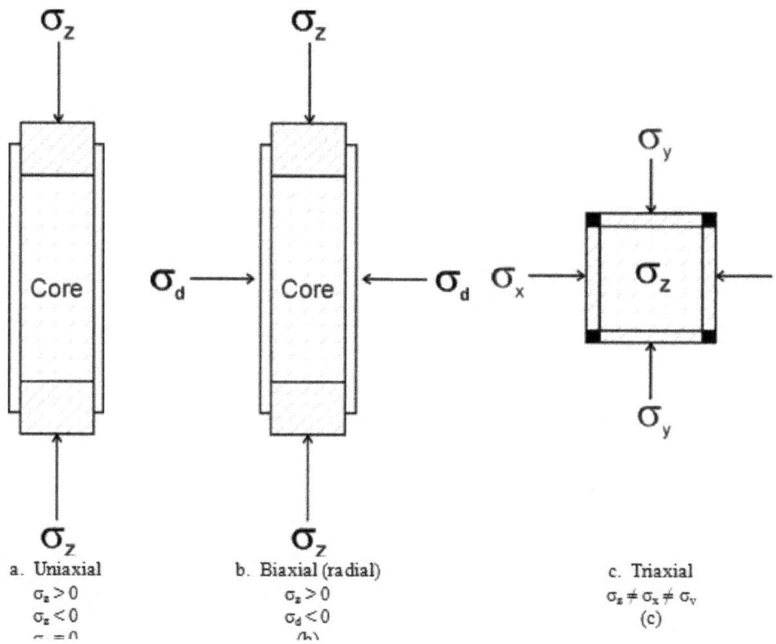


**Figure 3–5** *Schematic examples of uniaxial, biaxial and triaxial coreholders and tests.*

## 3.6 Effective Stress

Every location in the subsurface is subject to stresses from tectonic and overburden forces. These are exhibited by the deformation of structures into faults and folds over long periods of time. The stresses acting at any cube in the subsurface may be combined into three mutually perpendicular principal stresses that are unequal in magnitude (see Figure 3–6a and Figure 3–6b). One of the stresses is vertical ($\sigma_z$) and is due to the pressure produced by rocks and interstitial fluids resting on top of the cube; the other stresses are horizontal ($\sigma_x$ and $\sigma_y$, the resultants of stresses accumulated from the combination of tectonic and fluid forces); these are not equal, but because they cannot be accurately determined, they are assumed to be equal for practical computation.

The vertical stress ($\sigma_z$) of subsurface rocks is the integral of the density of the overburden rocks saturated with fluids:

$$\sigma_z = \frac{1}{144}\int_0^D \rho dD \tag{3.7}$$

A reasonable approximation of the overburden stress that is generally used is: 1.0–1.1 psi/ft of depth.

Porous subsurface rocks are generally saturated with water containing salts (brines), principally sodium and calcium chloride. Various saturations of gas, liquid hydrocarbons, and brine are found in porous sedimentary rocks. The fluids exert a pressure within the pores that has an effect on the overall mechanical behavior of the rocks. Thus, there are two stress components: (1) the skeletal *rock matrix*, and (2) the *pore pressure* of the fluids within the pores of the rocks.

In any subsurface formation containing fluids, the overall pressure of the fluids decreases the overburden pressure, thus the *effective* vertical stress of any fluid-filled subsurface formation is:

$$\sigma_z^* = \sigma_z - \alpha P_p \tag{3.8}$$

Experimental and field work have shown that the elasticity of the rock matrix reduces the effectiveness of the pore pressure by an amount equal to the *poro-elastic* constant ($\alpha$), which is a function of the ratio of the matrix compressibility to the bulk compressibility (Biot, 1955). Alpha theoretically ranges from zero to one and has an average value equal to 0.7 for reservoirs containing hydrocarbons.

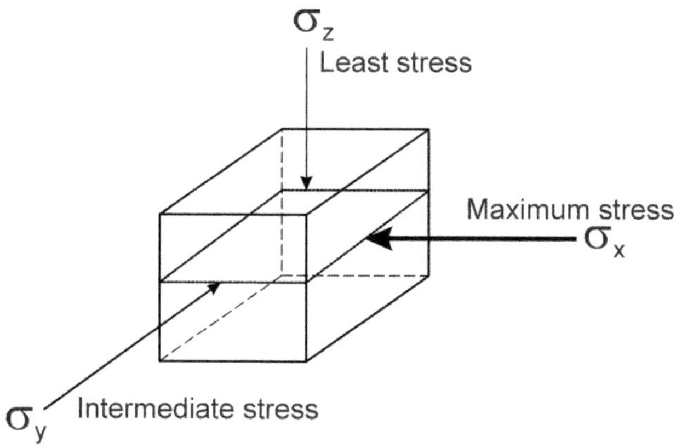


**Figure 3–6a** *Horizontal fracture occurs when the least principal stress is vertical $\sigma_x > \sigma_y > \sigma_y$ where the maximum stress is horizontal, a horizontal fracture will occur.*

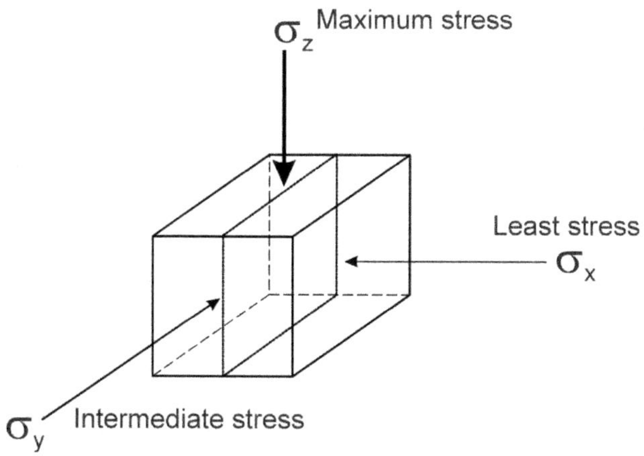


**Figure 3–6b** *Vertical fracture occurs when the least principal stresses are horizontal $\sigma_z > \sigma_y > \sigma_x$ where the maximum stress is vertical, a vertical fracture will occur.*

In accord with the elastic properties of rocks, which are expressed in terms of Young's modulus and Poisson's ratio (Eqs. 3.1 and 3.2), the other effective stress and strain relationships are generalized to three dimensions as follows (see Figure 3–7a):

$$\varepsilon_x = \frac{1}{E}\left[\sigma_x^* - \nu(\sigma_y^* + \sigma_z^*)\right] \tag{3.9}$$

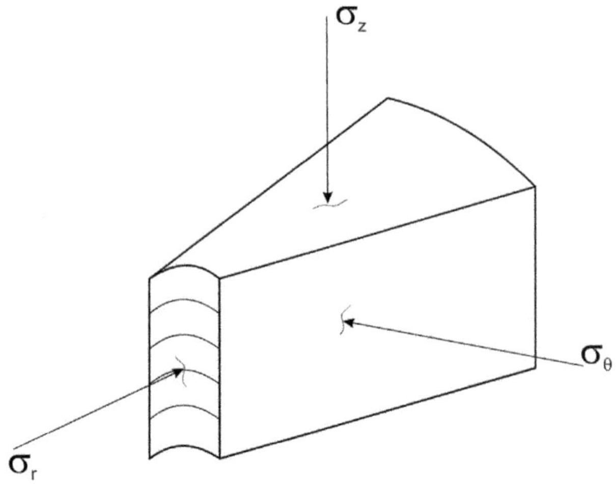


**Figure 3–7** *Segment of a cylindrical section about a wellbore illustrating the use of radial coordinates as defined by a set of equations.*

The assumption that tectonic forces are negligible is often made, in which case the horizontal stress in the x- and y-directions are equal. If the rock is further assumed to be laterally confined such that the horizontal strain is equal to zero ( $\varepsilon_x = \varepsilon_y = 0$ ), then the horizontal matrix stress becomes:

$$\varepsilon_x E = 0 = \sigma^*_{x,y} - \nu\sigma^*_{x,y} - \nu\sigma^*_z \tag{3.10}$$

$$\sigma^*_{x,y} = \left(\frac{\nu}{1-\nu}\right)\sigma^*_z$$

The *effective horizontal stress* acting at a point in a fluid-filled subsurface rock must include the pore pressure. The pore pressure acts to increase the horizontal, or radial, stress; expanding Eq. (3.10):

$$\sigma_x - \alpha P_P = \left(\frac{\nu}{1-\nu}\right)(\sigma_z - \alpha P_P) \tag{3.11}$$

$$\sigma_x = \left(\frac{\nu}{1-\nu}\right)(\sigma_z - \alpha P_p) + \alpha P_P$$

The general range of Poisson's ratio for sedimentary rocks is listed in Table 3–1. The overall range is from 0.15–0.30, which means that the horizontal matrix stress ranges from 18% to 43% of the vertical

**Table 3–1** Approximate Values of Poisson's Ratio for Sedimentary Rocks

| Hardness | Sandstone | Dolomite | Limestone |
|---|---|---|---|
| Hard | 0.15 | 0.25 | 0.23 |
| Medium | 0.17 | 0.27 | 0.25 |
| Soft | 0.20 | 0.30 | 0.28 |

stress, and thus the pressure required for propagation of a fracture is less than the stress of the overburden.

In an extreme case where the poro-elastic constant is equal to zero, pore pressure has no effect on the rock's mechanical properties. This is strictly theoretical because it means that the rock can expand elastically without resistance as pore pressure is increased. At the other extreme, where $\alpha = 1$, the pore pressure acts as a counteractive force against an applied force; in this theoretical condition the rock is completely rigid (it does not have any elasticity).

Geerstma (1953) used the ratio of the rock *matrix material compressibility* to the *bulk compressibility* to compute the value of the poro-elastic constant:

$$\alpha = 1.0 - \frac{C_r}{C_b} \tag{3.12}$$

An increase of fluid pressure will decrease the three compressive matrix stresses; continued increase of the pore pressure will eventually cause the stress with the least magnitude to reach a value equal to zero, and further increase of the pore pressure will produce a tensile stress in the direction of the least principal stress. If the tensile strength of the matrix is exceeded, the rock will rupture along a plane that is perpendicular to the least principal stress; when this occurs if the least principal stress is the vertical (overburden) stress, the rupture (or fracture) will be a *horizontal plane* through the cube as shown in Figure 3–6a ($\sigma_z \prec \sigma_{x\ or\ y}$;) and ($P_{BH} \succ \sigma_z + \sigma_T$). Thus, for a horizontal fracture to occur (using a non-penetrating fluid, due to fluid loss or formation damage near the wellbore), the bottom-hole pressure of fluid in the well adjacent to the formation must exceed the effective vertical stress plus the pore pressure and the vertical tensile stress of the formation:

$$P_{BH} \succ \sigma_z^* + P_p + (\sigma_T)_V \tag{3.13}$$

In geologic regions where structures are folded, or where thrust faults occur, the least principal stress is vertical, hence *horizontal fractures* are most probable ($\sigma_x > \sigma_y > \sigma_z$).

The conditions necessary to initiate a *vertical fracture* (with a non-penetrating fluid) require an additional increment of pressure; the bottom-hole pressure must exceed the relative strength of the two principal horizontal compressive stresses as well as the pore pressure and horizontal tensile strength of the rock.

$$P_{BH} \succ 3\sigma_x^* - \sigma_y^* + P_p + (\sigma_T)_E \qquad (3.14)$$

In areas where normal faults (normal to the Earth's surface) occur, the least principal stresses are horizontal and thus *vertical fractures* will occur ($\sigma_z > \sigma_x \geq \sigma_y$).

If a penetrating fluid is used, the pressure required to initiate and propagate the fracture are equal. A penetrating fluid increases the area of influence within the matrix of the rock and overcomes the compressive and tensile stresses at a lower pressure than required for a non-penetrating fluid.

## 3.7 Mohr Stress Diagram

Strain, stress, and rupture relationships can be determined using a cylindrical core, which is confined by an equal stress around the periphery ($\sigma_r$), and subjected to a compressive stress ($\sigma_z$) (see Figure 3–5) (Donaldson, et al., 1980). For rock cores at specific confining stresses, rupture will occur along a diagonal *shear plane* at a unique value of the confining stress; thus the value of the compressive stress, when rupture of the core occurs, is a function of the confining stress (see Figure 3–8).

The inherent disadvantage is that the rock sample is destroyed; therefore, several adjacent cores from a block of rock must be tested to obtain sufficient data for Mohr diagram analysis. At the instant of rupture, there is a shear stress along the shear plane ( $\sigma_\tau^*$ ) and a stress that is normal to the plane (see Figure 3–8). The angle ($\sigma^*_\tau$) between the shear plane and the confining horizontal stress ($\sigma_r$) is the angle of deformation, or angle of internal friction.

Using the effective shear stress, $\sigma_\tau^*$, and the stress normal to the shear stress, $\sigma_n^*$, as the coordinates of a graph (see Figure 3–9), the values of $\sigma_z^*$ and $\sigma_r^*$ that are recorded at the point of rupture of cores (I, II, and III) are plotted on a horizontal line as shown in Figure 3–9. A series of circles (whose diameters are equal to the compressive stress

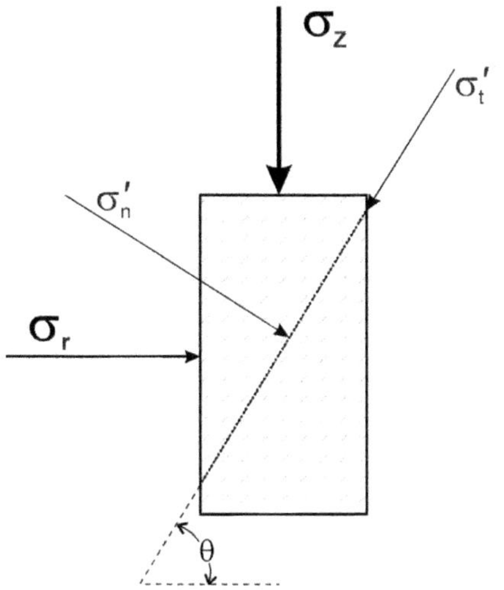


**Figure 3–8** *Rupture of the core will normally occur along a diagonal shear plane when subjected to a compressive stress ($\sigma_z$) under a specific confining stress ($\sigma$).*

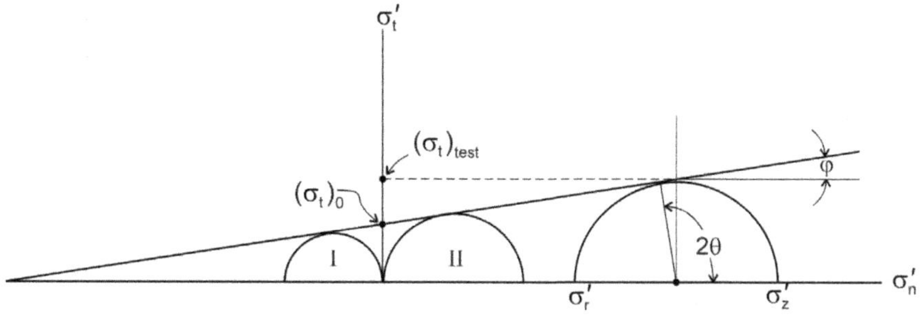


**Figure 3–9** *I Mohr circles of uniaxial tensile test ($\sigma_r = 0$).*
*II Mohr circle of uniaxial compressive test ($\sigma_r = 0$).*
*III Mohr circle of triaxial compressive test ($\sigma_r > 0$)*
$(\sigma_t)_{Test} = (\sigma_t)_0 + (\sigma_n)_{Test} * \tan\varphi$

minus the confining stress of each test) can be drawn on the graph. Lines tangent to the circles are drawn to a point on the horizontal axis; the area encompassed by the tangent lines is the *Mohr failure envelope*. Any combination of the two stressed ( $\sigma_z^*$ and $\sigma_r^*$ ) that falls within the envelope will not result in failure (rupture of the core); whereas, any combination that produces a circle whose apex is greater than the tangent line of the envelope will result in rupture (Tiab and Donaldson, 2004, 2012).

Several relationships are deduced from the Mohr diagram:

$$\text{Center of the circle} = \frac{\sigma_z^* + \sigma_r^*}{2} \; or \; \frac{\sigma_\theta^* + \sigma_r^*}{2} \tag{3.15a}$$

$$\text{Radius of the circle} = \frac{\sigma_z^* - \sigma_r^*}{2} \; or \; \frac{\sigma_\theta^* + \sigma_r^*}{2} \tag{3.15b}$$

$$\text{Normal Shear Stress, } \sigma_n^* = \frac{\sigma_z^* + \sigma_r^*}{2} + \left(\frac{\sigma_z^* - \sigma_r^*}{2}\right) cos(2\theta) \tag{3.15c}$$

$$\text{Shear Stress, } \sigma_\tau^* = \left(\frac{Diameter\ of\ Mohr\ circle}{2}\right) \sin(2\theta) \tag{3.15d}$$

$$\text{Angle of fracture, } \theta = 45° - 0.5\,\phi \tag{3.15e}$$

The fracture plane of a subsurface formation tends to extend along the axis of the principle stress, $\sigma_z^*$, or perpendicular to the direction of the least stress, $\sigma_r^*$. For example, where the horizontal, or radial, stress is greater than the vertical (overburden) stress, a horizontal fracture will occur (see Figure 3–6a). If the axis of the principal stress (in an isotropic matrix) is vertical, a fracture with a vertical plane will propagate from the point of applied stress (see Figure 3–6b).

There are subsurface conditions that affect the orientation of subsurface fractures: (1) internal rock stress (as a result of tectonic motions of rock structures), (2) differences of local rock properties due to variations of minerals, degree of grain cementation, variability of elastic properties, and (3) variation of pore pressure, etc. Such factors govern the orientation of the fracture planes and hence subsurface fracture plane orientation occurs at various angles from the horizontal axis.

Uniaxial (compressive and tensile), biaxial (radial), and triaxial tests are conducted on a series of adjacent cores obtained from a specific subsurface formation (when they are available) using laboratory equipment such as shown in Figure 3–5. The uniaxial tests are conducted with zero confining stress on the sides of the core; by convention, compressive stress is positive and tensile stress is negative. The triaxial tests use a core-holder that allows different amounts of stress to be applied on each side.

For the Mohr diagram, several triaxial tests may be conducted at increasing values of *confining stress,* which will normally increase the strength of the rock sample and thus displace the Mohr circle to higher values (see Figure 3–9). The test values obtained when the rock fractures, $\sigma_z'$, (the applied vertical stress on the cylinder of rock) and $\sigma_r'$ (the confining lateral stress) express the limiting diameters ($\sigma_z^* - \sigma_r^*$) of the Mohr circles.

Lines tangent to the circles (near the apex of the circle) are drawn to obtain the Mohr envelope (see Figure 3–9). The maximum shear stress value of each test is the vertical value at the tangent to the circle, and the angle between the failure envelope and the horizontal axis ($\varphi$) is the internal angle of friction. The maximum and minimum values of the normal stress occur at the intersection of the circle with the horizontal (normal axis). The intercept of the envelope and the vertical axis ($\tau_o$) is the cohesive strength of the rock.

The stress relationships of a well may be expressed in *radial coordinates* (see Figure 3–7) to represent the situation in the well filled with drilling mud (mud pressure = $P_\mu$) and the mechanical rock properties determined using a biaxial stress apparatus in the laboratory (see Figure 3–5) (Coats and Denoo, 1981):

$$
\begin{aligned}
\sigma_z &= \sigma_{ob} + 2\nu(\sigma_x - \sigma_y) \\
\sigma_{x,y} &= \left(\frac{\nu}{1-\nu}\right)\sigma_z \\
\sigma_\theta &= 3\sigma_x - \sigma_y - P_m \\
\sigma_r &= P_m
\end{aligned}
\tag{3.16}
$$

It is the *effective stress* that produces deformation of the rock skeleton: Hence, the equations above (Eq. (3.16)) are expressed in terms of the effective stress, and equal horizontal stresses $\left(\sigma_x = \sigma_y\right)$, when working with the Mohr diagram and field conditions:

$$
\begin{aligned}
\sigma_z^* &= \sigma_{ob} - \alpha P_p \\
\sigma_{x,y}^* &= \left(\frac{\nu}{1-\nu}\right)\sigma_{ob} + \alpha P_p\left(1 - \frac{\nu}{1-\nu}\right) \\
\sigma_\theta^* &= 2\sigma_{x,y} - P_m - \alpha P_p \\
\sigma_r^* &= P_m - \alpha P_p \\
\sigma_\tau^* &= 0.5(\sigma_z^* - \sigma_r^*)\sin(2\theta)
\end{aligned}
\tag{3.17}
$$

The imbalance in the subsurface between the *x* and *y* stress planes (common in many areas) is addressed by the multipliers using Poisson's ratio ($\nu$). Note that the pore pressure ($P_\pi$) and the overburden stress, $\sigma_z^*$, are constants; therefore, the mud pressure at the bottom of the borehole ($P_\mu$) is the only variable in the group of equations, Eq. (3.17). When $P_\mu$ is *increased,* $\sigma_r^*$ increases, and $\sigma_\theta^*$ decreases; when $P_\mu$ is *decreased,* $\sigma_r^*$ decreases and $\sigma_\theta^*$ increases.

Laboratory stress analyses establish the Mohr failure envelope that can be used to determine the maximum and minimum wellbore fluid pressure limits for injection and production operations to avoid formation damage, as illustrated in Example 3.2, using a Mohr diagram.

**Example 3–2** Example Calculations using the Mohr diagram

Assume that laboratory test on samples of a subsurface rock from a well yielded the following effective stresses (on Figure 3–10 as solid lines): (1) Tensile stress ($\sigma_\tau^*$) = –800 psi, (2) Uniaxial stress ($\sigma_z^*$) = 1,500 psi, (3) Radial stress ($\sigma_r^*$) = 2,500 psi, (4) Tangential stress ($\sigma_\theta^*$) = 6,000 psi, (5) Poro-elastic constant ($\alpha$) = 0.7, and (6) Poisson's ratio ($\nu$) = 0.25. Using these data, a Mohr diagram is drawn as shown in Figure 3–10 with the positive half of the Mohr circles, as solid lines. Sample calculations using hypothetical well data are described with dashed semi-circles.

The maximum shear stress of the largest circle ($\tau'$) occurs at the tangent of the Mohr circle, and also it may be calculated from the angles using Figures 3.15d and 3.15e.

$$\tan\phi = \frac{3}{10} = 0.300 \qquad \phi = 16^\circ\ 42'$$

$$\theta = 45^\circ - 0.5\phi, \qquad \theta = 36^\circ\ 39'$$

$$\sigma_\tau^* = \left(\frac{\sigma_\theta^* - \sigma_r^*}{2}\right)\sin(2\theta) = 1750(0.9578) = 1{,}676\,psi$$

**Well conditions I.** (Dashed line circle I, see Figure 3–10). The lowest stress state occurs at $P_\mu = P_\pi$. This is a mud weight gradient equal to the pore pressure.

1. Depth ($D$) = 8,000 ft
2. Overburden stress ($\sigma_{ob}$ = 1.1 psi/ft) = 8,800 psi

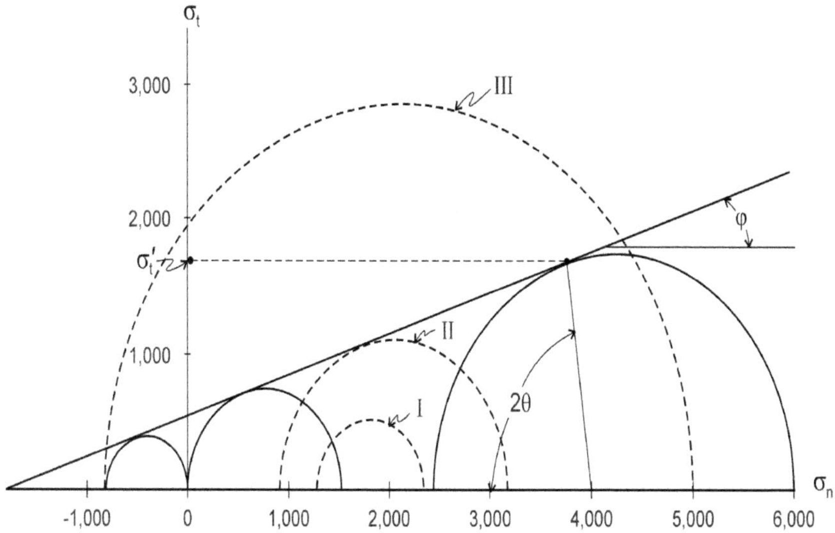


**Figure 3–10** *Mohr Circle example calculations: Semi-circles from laboratory data are solid lines that define the Mohr failure line for the formation rock. The dashed semi-circles represent the conditions expressed by examples I, II, and III.*

3. Pore pressure ($P_\pi = 0.46$ psi/ft) = 3,680 psi

4. Mud column pressure ($P_\mu = 0.46$ psi/ft) = 3,680 psi

$$\sigma_z = 8{,}800 - 0.7(3{,}680) = 6{,}224$$

$$\sigma_{xy} = 0.333(8{,}800) + 0.7(3{,}680)(0.667) = 4{,}651$$

$$\sigma_r^* = 3{,}680 - 0.7(3{,}680) = 1{,}104$$

$$\sigma_\theta^* = 2(4{,}651) - 3{,}680 - 0.7(3{,}680) = 3{,}046$$

$$\sigma_\tau = \left(\frac{\sigma_\theta - \sigma_r}{2}\right)\sin(2\theta) = \left(\frac{3046 - 1104}{2}\right)(0.9578) = 529$$

The circle is within the envelope and the formation is stable.

**Well conditions II.** (Dashed line circle II, see Figure 3–10). Shear failure occurs when the circle touches, or exceeds, the envelope line. When this condition exists, wash-out will occur in the softer sands and shales during drilling and sanding will occur during production draw-down.

Calculate the point at which shear failure will occur by trial and error. A decrease of mud pressure will cause a decrease of the radial stress and increases of the tangential stress, thus expanding the size of the circle until it touches the envelope line.

Reduce $P_\mu$ to 3,550 psi (gradient – 0.444 psi/ft)

$$\sigma_r^* = 3550 - 2576 = 974$$

$$\sigma_\theta^* = 9302 - 3550 - 2576 = 3176$$

$$\sigma_\tau = \left(\frac{3176 - 974}{2}\right)(0.9578) = 1055\,psi$$

**Well conditions III.** (Dashed line circle III, see Figure 3–10). Tensile failure will occur if the tangential stress becomes equal to or greater than the tensile strength of the rock ( $\sigma_T^* = -800$ ).

$$\sigma_\theta^* = -800 = 2(4651) - P_m - 0.7(3680)$$

$$P_m = 7526 \qquad Gradient = \frac{7526}{8000} = 0.9407\,psi/ft$$

$$\sigma_r^* = 7526 - 0.7(3680) = 4950$$

$$\sigma_\tau^* = \left(\frac{4950 - (-800)}{2}\right)(0.9578) = 2754$$

The shear stress is above the Mohr envelope and failure will occur, but in this case the area immediately around the bottom-hole is in tension, and failure is due to increased tensile stress produced by the large increase of fluid pressure in the pores. Note that the positions of $\sigma_\theta^*$ and $\sigma_r^*$ are reversed from those of compressive stress in the example.

---

Fracture (or failure) of the rock at the bottom of the well occurs when the stress on the skeletal structure of the rock reaches critical values of the effective stresses acting on the rock, as defined by the Mohr stress envelope (see Figures 3–10, 3–11). The stress characteristics of a specific well $(\sigma_{ob}, \sigma_{x,y}, P_p, \alpha, \text{and } \nu)$ are constants; therefore, the only variable controlling the radial and tangential stresses is the pressure of the mud column at the bottom of the well.

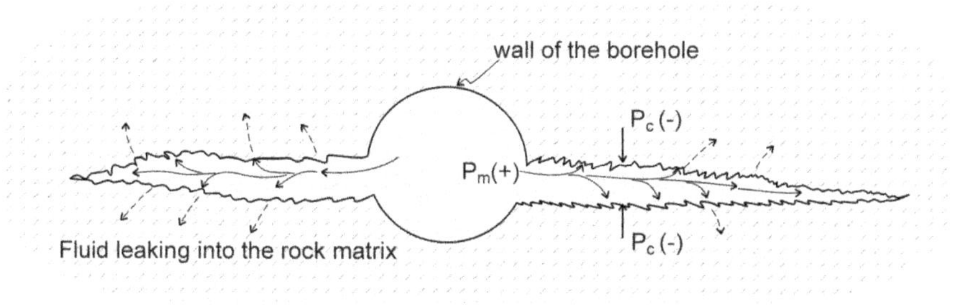


**Figure 3–11** *The mud pressure at the bottom of the hole propagates the fracture when it is positive and greater than the fracture closure pressure which is a negative value acting to close the fracture.*

Any combination of values of $\sigma_\theta^*$ and $\sigma_r^*$ (due to increase or decrease of the mud pressure $P_\mu$) that exceed the shear stress at the edge of the Mohr envelope will result in failure, which is either excessive compressive stress, or tensile stress, within the pore matrix structure that tears the rock apart.

In the subsurface, the three stresses are generally unequal. In tectonically relaxed areas, the stress with the greatest magnitude is the vertical (overburden) stress. In active tectonic regions, however, the three stresses can have considerably different values. Natural faults occur along the positive line of the Mohr envelope and have a fault plane that is usually at an angle equal to 45° – $\phi$/2, with respect to the least stress.

After a fracture is initiated, fluid entering the fracture applies pressure to both side of the fracture, pushing them apart and advancing the fracture (see Figure 3–11). The amount of pressure required to hold the fracture open is equal to the natural horizontal stress acting perpendicular to the fracture plane, or closure pressure ($P_{\chi\lambda}$), (see Figure 3–11). The fracture will continue to increase in length so long as the pressure exerted near the wellbore is greater than the inter-granular strength of the rock. There is a pressure drop between the leading edge of the fracture (at the wellbore) and the tip of the advancing fracture due to fluid friction loss along the walls of the fracture. Fracture growth will end when the combination of pressure losses from (1) fluid loss into the matrix, (2) friction in the pipe, and (3) friction on the wall of the fracture are equal to the overall closure pressure.

## 3.8 Initiation of Fractures

The process of fracture initiation and propagation, considering the width, height, and length of the fracture is very complex; it must be simulated by solution of partial differential equations using a computer to obtain accurate predictions of fracture propagation. The injection of a fluid to create, and propagate, the fracture changes the stress distribution within the formation and creates a long, narrow fracture extending from the wellbore; are functions of the fracture properties, rate and pressure of injection, fluid density, viscosity and reactivity (acidic) of the frac-fluids. In addition, some of the injected fluid flows into the un-fractured areas of the formation (called leak-off), which changes the temperature and thus the matrix rock properties; and if an acid is used in the injection fluid, it will modify the natural stress characteristics.

Subsurface rocks are under compressive stress due to the mass of the overburden stress (approximately 1.0 psi/foot of depth); this produces an effective compressive stress within the subsurface formation. The effective compressive stress can be estimated as follows:

$$\sigma^*_{x,y} = \left(\frac{\nu}{1-\nu}\right)\left(\sigma_{ob} - P_p\right) \tag{3.18}$$

Poisson's ratio for limestone in the mid-continent of the United States has an average value of 0.27, sandstone is usually around 0.32, and shale is about 0.26. As a consequence of the relationship expressed in Eq. (3.18) and the values of Poisson's ratio, the horizontal stress in the matrix generally ranges from one-third to half of the net vertical, overburden stress. Ridged reservoirs composed of strongly cemented sands (where the cementing agent is a compound such as siderite, $FeCO_3$), limestone or dolomite, counterintuitively require lower pressure for fracture initiation than softer formations such as those composed of unconsolidated sand and shales. Consider Example 3-3.

**Example 3–3**

Assume an overburden stress of 6,000 psi and pore pressure of 2,000 psi.

*Limestone* formation, Poisson's ratio = 0.27.

$$\sigma^{*}_{x,y} = \left(\frac{0.27}{1-0.27}\right)(6{,}000 - 2000) = 1.480\,psi$$

Ratio of horizontal stress to the net vertical stress =

$$\frac{1{,}480}{4{,}000} = 0.37$$

*Shale* formation, Poisson's ratio = 0.36

$$\sigma^{*}_{x,y} = \left(\frac{0.36}{1-0.36}\right)(6{,}000 - 2{,}000) = 2{,}250\ psi$$

Ratio of horizontal stress to the net vertical stress =

$$\frac{2{,}250}{4000} = 0.56$$

Note that the limestone requires 1,480 psi for fracture initiation, whereas the softer shale requires 2,250 psi (almost 1.5 times greater).

---

The pressure of the fluids in the reservoir (water, oil, gas) has a significant effect on the horizontal stress, in accord with Eq. (3.18). An increase of the pore pressure ($P_{\pi}$) decreases the horizontal stress; therefore, a fracture initiation fluid that *penetrates* the matrix around the wellbore will *initiate* and *extend* the fracture away from the wellbore at a constant pressure (see Figure 3–12a). Whereas a frac-fluid that does not penetrate into the matrix (because of high viscosity, very low permeability ($<1.0$ md), or formation permeability damage in the zone around the wellbore) requires an additional initial breakdown and increment of pressure in order to initiate the fracture (see Figure 3–12b).

Some of the injected fluid flows into the un-fractured areas of the formation as a function of its permeability and porosity (called *leak-*

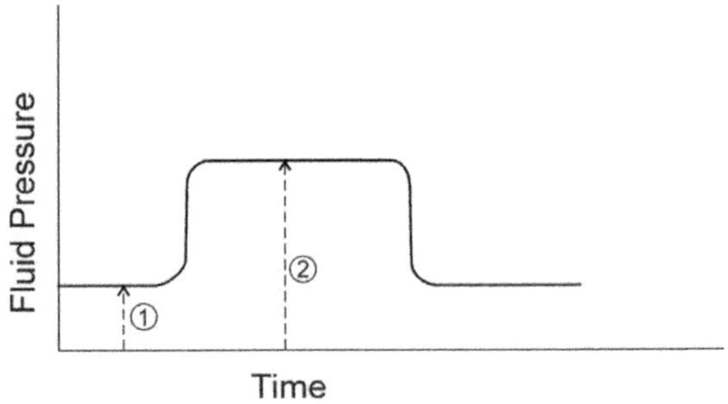


**Figure 3–12a** *Fluid pressure in the wellbore (BHP) just before, during, and after fracture initiation with a matrix penetrating frac-fluid: (1) natural reservoir fluid pressure and (2) the fracture initiation and extension fluid pressure.*

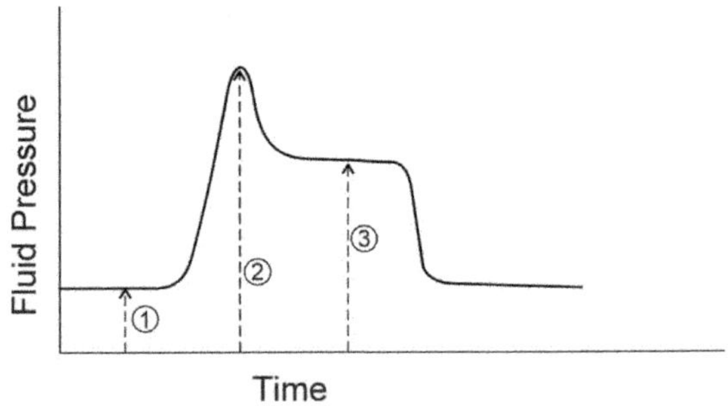


**Figure 3–12b** *Fluid pressure in the wellbore (BHP) just before, during, and after fracture initiation with a non-penetrating frac-fluid: (1) natural reservoir fluid pressure, (2) fracture initiation (or break-down pressure) and, (3) fracture extension (or propagating pressure).*

*off*); this changes the temperature and thus modifies the natural stress characteristics of the matrix rock properties (see Eq. (3.18)).

As shown in Figure 3–14b, after a short duration, peak pressure breaks the face of the rock at the wellbore and initiates the fracture. A slightly lower pressure of frac-fluid entering the fracture is required to extend (or propagate) the fracture. This *fracture propagation pressure* (synonymous with the bottom-hole pressure) must be greater than the sum of: (1) the pore pressure, (2) the minimum effective horizontal stress, (3) the pressure loss to friction from fluid flow against the expanding walls of the fracture ($P_{\phi\rho\iota\,\chi}$), and (4) pressure loss due to leak-off of the frac-fluid ($P_{\Lambda O}$):

$$P_{prop} = (P_p + \sigma_\theta^*) + P_{fric} + P_{LO} \tag{3.19}$$

$$P_c = P_{prop(P-BH)} - P_{res} \tag{3.20}$$

where

$P_{\pi\rho o\pi(\Pi\text{-}BH)}$ = fracture propagation pressure (bottom-hole pressure)

$P_p$ = formation pore fluid pressure

$P_{\phi\rho\iota\chi}$ = pressure loss due to friction in the pipe

$P_{\Lambda O}$ = pressure loss due to leak off

$P_\chi$ = fracture closure pressure

$P_{\rho\epsilon\sigma}$ = static reservoir pressure

The *fracture closure pressure* is the pressure that the proppants must counteract to hold the fracture open when the fracture treatment has been completed. Closure pressure is equal to the pore pressure plus the effective horizontal reservoir pressure.

The net fracture pressure acts against the elasticity of the rock (Young's modulus of elasticity) to increase the width of the fracture. The behavior of the net fracture pressure during fracture operation may be used for analysis of the fracture extension progress. A plot of the Log($P_{\nu\epsilon\tau}$) versus Log(*time*) yields straight lines that are indicative of the fracture dynamics (see Figure 3–13), (1) when Log($P_{\nu\epsilon\tau}$) increases linearly with respect to Log(*time*) after initiation of the fracture, and the fracture is extending laterally away from the wellbore with minimum vertical extension (assuming that impermeable barriers exist above and below the formation); (2) if the Log($P_{\nu\epsilon\tau}$) remains constant, an abnormal condition has developed due to intersection of the growing fracture with a high permeability zone such as a natural fault causing the fracture growth to stop; (3) if the Log($P_{\nu\epsilon\tau}$) suddenly increases abnormally, fracture growth has stopped, possibly because an impermeable barrier was encountered, and the fracture width at the obstruction is increasing (ballooning against the obstruction); (4) if the Log($P_{\nu\epsilon\tau}$) decreases sharply, it probably means that the fracture has broken through one of the confining barrier zones above or below the formation. In addition, fracturing changes the flow pattern of the formation fluids flowing into the well. When the well consists of a circular borehole, a radial flow pattern exists in which the flow lines of

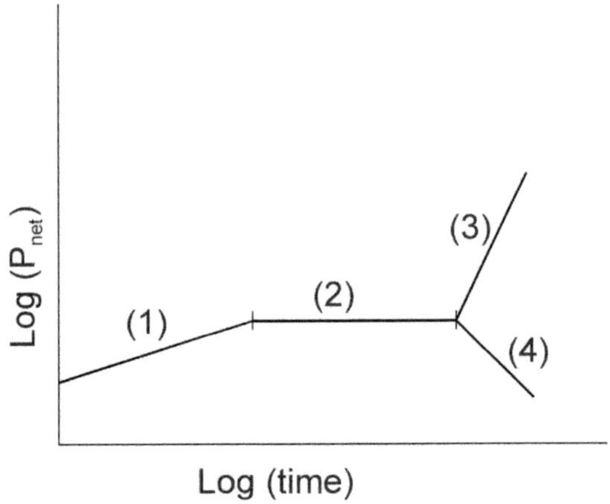


**Figure 3–13** *Behavior of the Log ($P_{net}$) during fracture operation: (1) fracture increasing in length after initiation, (2) Log ($P_{net}$) constant if a high permeability zone has been encountered, (3) sharp increase when an impermeable zone is encountered, (4) decrease if the fracture breaks through a confining formation above or below the formation being fractured.*

fluids entering the well are radially oriented (see Figure 3–14a). However, when a fracture system is associated with the well, the pattern of flow changes to one that is linear; formation fluids flow first into the fracture and from the fracture into the well (see Figure 3–14b). Hence, after completion of the fracture, the fluid conductivity should be great enough to eliminate the radial flow pattern; in other words, fracture permeability should be several orders of magnitude greater than the matrix permeability.

Computer programs are required to include the numerous variables associated with fracture mechanics to ascertain reasonably accurate prediction of fracture dimensions. However, a simplified expression for fracture width ($W$) is used for estimation of the influence of the major parameters involved (Geertsma and de Klerk, 1969):

$$W = C\left(\frac{Q\mu R}{E}\right)^{1/4} \tag{3.21}$$

Eq. (3.21) shows that fracture width is a function of the fourth root of four variables. Thus, variations of each variable must be large in order to have an effective influence on the width of the fracture. The injection rate ($Q$) and fracture radius can only be changed by about ten times, so their influences are minimal. Young's modulus is a

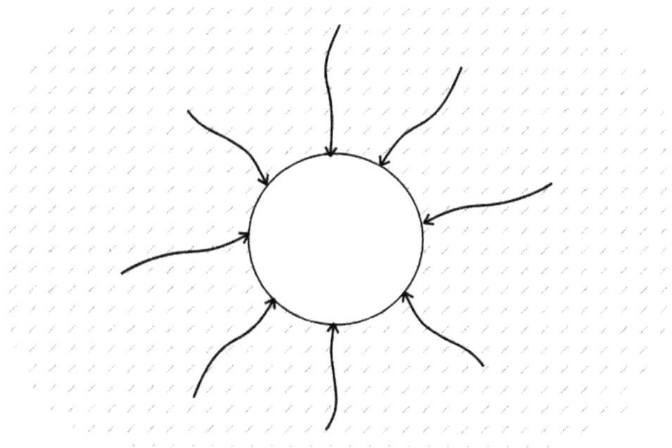

**Figure 3–14a** *Radial flow of formation fluids into an un-fractured well.*

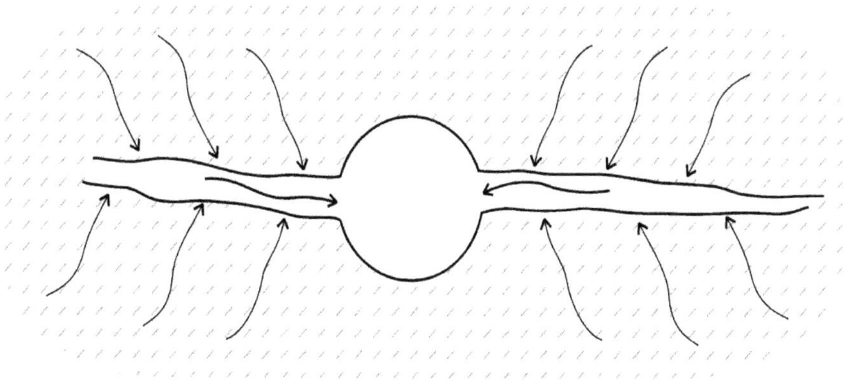

**Figure 3–14b** *Linear flow of formations into a fracture followed by linear flow into the wellbore.*

set property of the formation rock; however, the viscosity of the frac-fluid can be changed by as much as several hundred times greater than the viscosity of water (1.0 *cP* at 68°F); the higher viscosities can induce fracture widths 5–6 times greater than is possible with water, thus making viscosity the most important variable.

## 3.9 Propping the Fracture Open

The objective of well productivity stimulation by hydraulic fracturing is to obtain and maintain an increased rate of hydrocarbon productivity from an oil or gas well. The fractures result in an increase of global permeability in the vicinity of the wellbore that is thousands of times greater than the matrix permeability. However, if the width of the fracture is not supported against the natural horizontal stresses that create the closure pressure, the fracture will close and heal itself by sealing fluid flow channels. Therefore to accomplish this task, a variety of propping agents have been developed to meet the demands of numerous subsurface fracture environments. The important criteria that must be met are:

*Strength*—The load required to crush a single grain divided by the diameter-squared of the contact at the point of crushing. The proppant must be strong enough to withstand the fracture closure stress ($P_{\chi\lambda}$); if not, it will be crushed and the fracture will close causing a drastic decrease of fracture permeability. Sand can be used for $P_{\chi} < 6{,}000$ psi; intermediate-strength proppants, for $P_{\chi\lambda}$ for 5,000–10,000 psi; and high-strength proppants for $P_{\chi\lambda} > 10{,}000$ psi.

In accord with Eq. (3.19), when the bottom-hole pressure is greater than the reservoir pressure, the numerical value of the closure pressure is positive; therefore the frac-fluid pressure is propagating the fracture. When $P_{\chi\lambda} = 0$, the propagation pressure is equal to the reservoir pressure and propagation of the fracture has ceased; the fracture is being held open by the frac-fluid pressure. When the bottom-hole fluid pressure is negative during production when the well is pumped, the negative value of the bottom-hole augments the closure pressure; pressure is exerted against the proppants in the fracture that can cause crushing and embedment of the proppant (see Figure 3–15).

A lower pressure at the wellbore during production results in an increase of the closure pressure in the localized zone of the wellbore that can cause crushing and/or embedment of the proppant into the walls of the fracture. Hence, the area near the wellbore is the most critical for proppant placement, especially for a formation with a relatively soft matrix. A soft matrix requires placement of multiple layers (at least five layers) of proppant in the near-wellbore area. Proppant concentrations of at least 500 lbs/1000 ft$^2$ are required to achieve multilayered proppant placement. For a formation with a hard matrix, where embedment of the proppant does not take place, the fracture opening can be supported by a monolayer of proppant.

*Self-propping*—In formations that have acid soluble components, an acidic solution is used as the frac-fluid to create fluid-flow channels within the walls of the fracture. These etched channels remain in place when the well is placed on production; thus propping agents are not required.

*Fluid viscosity and proppant density*—Fluid viscosity and the density of the proppant are the most critical factors controlling the rate of settling of the proppant from the frac-fluid and placement, or accumula-

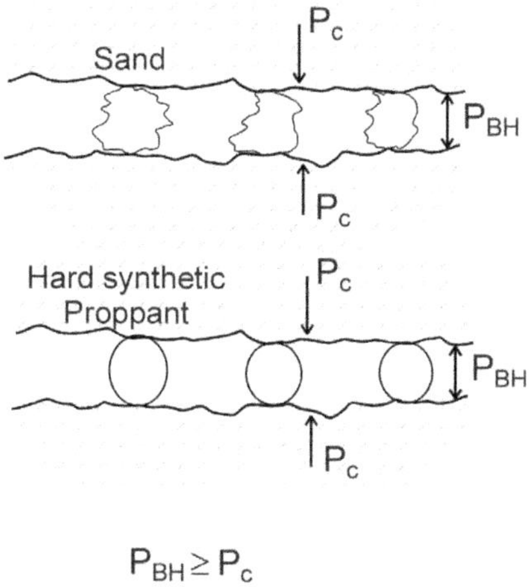


**Figure 3–15** *Effect of frac-fluid ($P_{BH}$) pressure changes on proppants. The closure pressure can be great enough to crush sand grains.*

tion, of the proppant in the fracture. Laboratory experiments and field results have elucidated the patterns of proppant placement associated with low and high viscosity fluids. On entering a fracture, sand suspended in a *low viscosity* fluid immediately begins to precipitate near the wellbore and builds up to an equilibrium height, which is a function of viscosity of the fluid and density of the proppant (Figure 3–16a). As the suspension continues to enter the fracture, additional sand, dropping out of the suspension, rolls across the top of the initial bank of sand and deposits farther out in the fracture. As the suspension continues to be pumped into the fracture, the rate of sand deposition increases as the viscosity of the fluid decreases from an increase of temperature and time while the velocity simultaneously decreases from fluid loss to the formation matrix.

When a *high viscosity* sand suspension is used, the sand bank builds up slowly and extends well into the fracture because the settling rate of the sand is comparatively slow with respect to the low viscosity fluid (see Figure 3–16b). At completion, the proppant placement in the fracture occupies a greater percentage of the height and length of the fracture, since viscosity of the frac-fluid is greater and leak-off is much less.

*Type and size of proppants*—Fracture conductivity depends to a large extent on the type, uniformity and size of the proppants, amount of embedding into the matrix, crushing of the proppant, and placement in the fracture. The proppant must be refined for uniformity in size because fine particles will block flow channels and

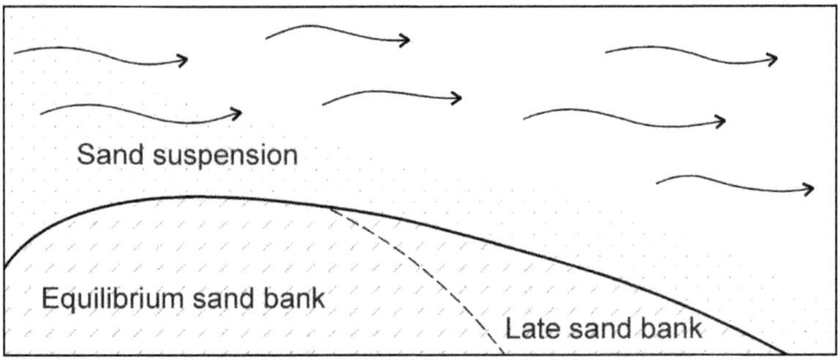


**Figure 3–16a** *General performance of a low viscosity fluid sand suspension entering a fracture. Sand falls out quickly near the wellbore.*

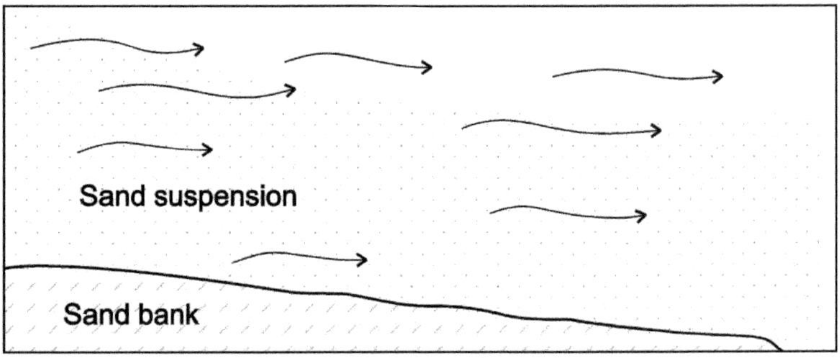


**Figure 3–16b** *General performance of a high viscosity fluid sand suspension entering a fracture. A sand bank builds up slowly and extends far into the fracture.*

seriously decrease the permeability of the fracture. Synthetic proppants of various types are used because their size distribution can be controlled during manufacture and their density (and hence settling rate) can be varied.

Table 3–2 illustrates the size and approximate permeability and crushing strength of three popular sand sizes. The sands are mined from outcrop sandstone formations, some in Illinois (Ottawa sand) and Texas (Hickory sand); The sand is crushed, screened, and washed to secure a clean, uniform sand with an average density of 2.7 g/cm$^3$. Polymer (or resin) coated sand is more uniform in size and can withstand higher closure pressures for use in deep formation. When very high closure pressures require hard proppants, sintered bauxite, aluminum oxide (corundum), or zirconium oxide with densities greater than 3 are used.

**Table 3–2** Approximate Ranges of Proppant Properties

| Mesh size | Diameter, inches | Pack permeability | Crushing strength |
|---|---|---|---|
| 20/40 | 0.033–0.016 | 120 | 10–12,000 |
| 10/20 | 0.079–0.033 | 200 | 8–10,000 |
| 8/12 | 0.094–0.066 | 1,500 | 4–6,000 |

Large size (4–8 mesh; 0.02–0.09 inch) provide greater fracture permeability but are more difficult to transport and place properly. In addition, sandstone formations that are subject to fines migration are prone to plugging of the pore channels with resulting drastic reduction of permeability. Smaller proppants that resist invasion and bridging of pore by fines are much more suitable for use in formations that are subject to fines migration. Use of smaller proppants under these conditions will yield lower permeability, and hence, lower initial conductivity of production fluids. This means that over the lifetime of the well, the overall productivity will be much greater. Also, large grain sizes are more susceptible to crushing and therefore cannot be used in deep wells (>6,000 feet deep). An approximate guide to proppant selection based on closure pressure is as follows: sand for a closure pressure up to 6,000 psi; resin-coated sand up to 8,000 psi; intermediate-strength ceramic from 3,000–10,000 psi; bauxite from 5,000–20,000, and high-strength bauxite from 15,000–25,000 psi.

*Bulk quality*—Considerable care is taken to remove fines and impurities from sands. Impurities include feldspars, carbonates, and iron oxides; a maximum of such impurities is allowed in any proppant. Included in the quality is the roundness (sharpness of corners and grain curvature). Stress from the closure pressure is more evenly distributed on smooth, spherical grains and can thus support greater pressure.

Relatively low density decreases precipitation from the frac-fluid and promotes deeper and more uniform placement. High-density proppants are difficult to maintain in suspension and transport within the fracture. Use of high viscosity fluids is required to move the high density proppants over the full length of the fracture; use of low viscosity fluids (see Table 3–3) requires high injection rates.

**Table 3–3** Use of High and Low Viscosity Fluids to Move Proppants

| Property | Size-sieve mesh | Density | Fluid Viscosity | Injection Rate |
|---|---|---|---|---|
| High/Large | 6/12 | 140 $lbs/ft^3$ | 180 cP at 60°F | 20 bbl/min. |
| Low/Small | 70/140 | 100 $lbs/ft^3$ | 1 cP at 60°F | 5 bbl/min. |

CHAPTER 4

# Fracture Fluids

## 4.1 Introduction

The objective of hydraulic fracture well treatment is simply to stimulate the production of hydrocarbons from a subsurface geologic formation. The successful production of gas from shale reservoirs around the world is attributed to the development of horizontal drilling techniques, innovative formulations of fluids to create extensive fractures within the shale formation, and real-time micro-seismic observation of the fraction network development.

Fracturing fluids are complex mixtures containing as many as six or seven different components. Many low concentrations of various additives are used to control specific behavioral characteristics of the frac-fluids at several distinct phases of the fracture treatment. The relative compatibility of the mixed additives must be determined for each formulation to avoid reactions that will lead to loss of the reacting additives through precipitation and/or deactivation.

Several methods have been developed to accomplish the task of fracturing formations from horizontal wells that can extend a half-mile, or more, away from a vertically drilled well from which the horizontal branches radiate laterally. Figure 4–1 is an illustration of one to the techniques that are used. A tube containing multiple movable packers with sleeves that initially cover the perforations have a funnel shape with outlet holes of varying sizes, from large to small. The packers divide the horizontal well-tubing into sections (see Figure 4–1). The section at the end of the tubing is the first to be fractured After the proppant is introduced, it is followed by a ball which is small enough to pass through all but the last packer, where it lodges and seals that section. The increased pressure exerted on the movable packer causes it

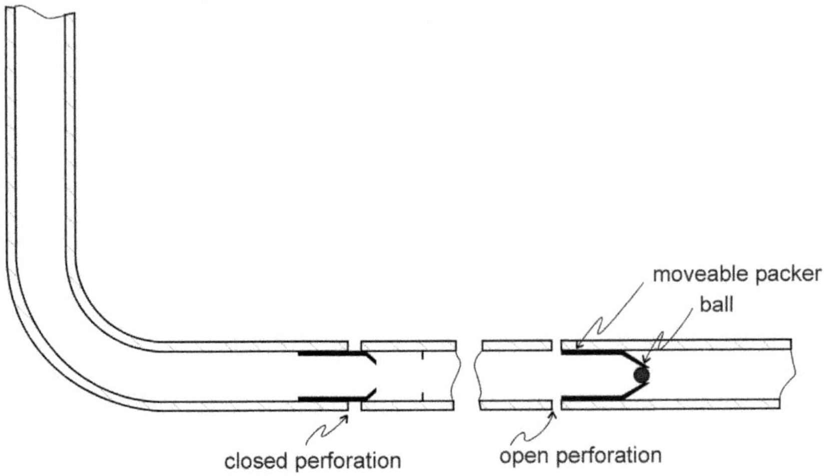


**Figure 4–1** *After placing the proppant, a ball larger than the movable packer's opening is inserted. The high pressure behind the packer causes it to move and seal the section in front of it while opening the perforations in the section behind it.*

to move and uncover the perforations of the section behind it which is fractured next, and so on until the final section next to the vertical drill hole is fractured. The blocked packers are then drilled out for final completion of the fracture treatment.

To accomplish the task of fracturing the formation, the fluid used to create the fractures in the oil or gas reservoir (which will effectively increase the area of the wellbore) must have a large number of unique attributes. This requires a thorough understanding of the complete rheological properties of the mixture of chemical compounds which make up the frac-fluid (and its proppants) that is designed to accomplish the task of creating a network of fractures extending away from the wellbore and is interlaced with the natural fracture system of the shale reservoir. For example: if high viscosity water-gel is selected, the minimum amount of gel-producing agent should be used with inclusion of a gel-breaking reagent that will be effective at a temperature which is about 20% lower than the overall reservoir temperature. It should contain a second gel-breaker designed to be effective for the back-flow frac-fluid (at a temperature of 10–20°F greater than ambient temperature) which will be accumulated in a tank at the end of the fracture treatment. In addition, although fresh water frac-fluid is desirable because of its availability, the water sensitivity of the clays within the shale must be considered. A minimum amount of potassium chloride (*KCl*) must be added to protect against the clays' hydration (swelling by adsorption of water) that can ruin the shale matrix permeability to gas. A surfactant suitable for microemulsion generation may be added to enhance the proppant carrying capacity of the frac-fluid, but the surfactant must be compatible with the reservoir brine (not form precipitates). A biocide must undoubtedly be used to control aerobic bacteria in the fluids in mixing tanks and the detrimental introduction of bacteria that can create long-term ruinous reservoir problems, such as production of hydrogen sulfide throughout the reservoir. All of these compounds are mixed together to form the frac-fluid that must:

- Initiate and extend a cleavage, or fracture, of the reservoir rock using a minimum level of pressure,
- Suspend and transport particles, such as graded sand, deep into the fracture to hold it open for drainage of hydrocarbons from the formation to the wellbore,
- Not leak into the formation matrix, displacing or compressing the reservoir fluids,
- Exhibit low pressure-loss in the pipe due to friction. The velocity of the fluid in the fracture greatly determines how the proppant settles in the fracture as it is carried away into the wellbore,
- Be compatible with the formation minerals and fluids (not cause precipitates),
- Not damage the formation matrix permeability by establishing fluid blocks, particle transport, precipitation, pore plugging by transport of fine particles, and
- Be easy to remove from the fracture after that task has been completed.

Subsurface porous formations containing oil and/or gas exhibit remarkable variation in chemical and physical properties, such as rock composition, permeability and porosity, temperature and pressure, interstitial fluid composition, etc. Therefore, the frac–fluids must be tailored to meet the unique conditions of each reservoir, and frac-fluids with many different physical and chemical properties have been developed by addition of chemical compounds to oil-and water-based frac-fluids.

The process of fracture treatment may be described very briefly, and generally, as a set of stages:

*Stage 1. Fracture initiation*—This is accomplished by pumping fluid at a rate faster than the leak-off rate into the formation in an open hole at the bottom of the well, or through perforations in the casing. Fluid pressure is increased in the rock to overcome the compressive stress of the rock. The rock then breaks (fractures) along a plane perpendicular to the minimum compressive stress in the matrix. After parting, the fracture is extended in width and length as fluid pressure in the fracture works against the elasticity of the rock,

*Stage 2. Proppant placement*—After the fracture is widened sufficiently to accept proppant material, sand or other granular matter is added to the frac-fluid and pumped into the fracture. Then, the fracture grows upward, downward, and outward,

*Stage 3. Ending of fracture growth*—When the rate of frac-fluid leak-off equals the rate of fluid injection. When enough proppant has been added, pumping is stopped and the pressure in the fracture decreases. Growth of the fracture may also end due to "sand-out" As sand deposits in the fracture, greater pressure is required to increase the fracture length. If the limit of pressure application has been reached, fluid injection slows, and may stop, causing sand to drop out of suspension and thus ending the fracture treatment prematurely,

*Stage 4. Removal of frac-fluid*—After a rest period to allow time for the viscosity breakers to decrease the viscosity of the frac-fluid in the fracture, the well is pumped to remove the frac-fluid and fluid loss additives and to place the well back into production at a considerably enhanced flow rate.

## 4.2 Oil-Based Fracturing Fluids

The first frac-fluids were light crude oil (API° >35°), gasoline, kerosene, and diesel oil. The process was first initiated by pumping produced fluids back into the well at high pressure until a sudden pressure-drop indicated that fracture of the rock around the wellbore had been initiated. Next, an oil-based mixture containing sand (to hold the fracture open) would be injected and then the well was placed on production. The oil-based frac-fluid would be produced from the fracture and blended into the produced formation hydrocarbons.

Oil-based mixtures were used because they were believed to be less damaging (reduction of the reservoir rock permeability) than water-based fluids. They were used in cold weather when water-based fluids would be subject to freezing. It was quickly recognized that prevention of frac-fluid loss by leak-off into the formation rock matrix and better suspension and placement of the frac-sand could be accomplished by somehow increasing the viscosity of the frac-fluid.

The efficiency of the frac-fluid in creating the fracture is based on having a low *leak-off rate*, or a loss of frac-fluid from the fracture by displacement of reservoir fluids as the fracture increases in height and length.

Perhaps the first high viscosity (thickened) fluid was gasoline that was formed into a gel by addition of napalm (aluminum soaps of carboxylic acids such as aluminum- octoate–Appendix B). Napalm also was then generally used to thicken the other oil-based frac-fluids. Sometime in the 1970's napalm was replaced by aluminum-phosphate-ester which enhanced the temperature stability of the gel as well as the proppant carrying ability. Aluminum hydroxyl complexes, associated with loose covalent bonding with the phosphate ester, form long-chain polymer molecules that produce the gel by incorporating the hydrocarbon oil (see Figure 4–2). Sodium aluminate, frequently used in water treatment for domestic and industrial use, may be added to enhance the reactivity of the phosphate ester. The aluminum fluid is proportionally metered into the well tubing as the oil is being pumped; thus, the gel is formed by turbulent mixing in the tube. The amount of gelling agent is between 5 and 10 gallons/1,000-gal of oil, yielding an apparent viscosity of 150–200 *cP* at 175 $s^{-1}$, and can be used with a bottom hole temperature (BHT) up to 250°. At the end of the fracture treatment, the viscosity of the frac-fluid is reduced by the addition of a weak acid or base that interrupts the covalent bonding.

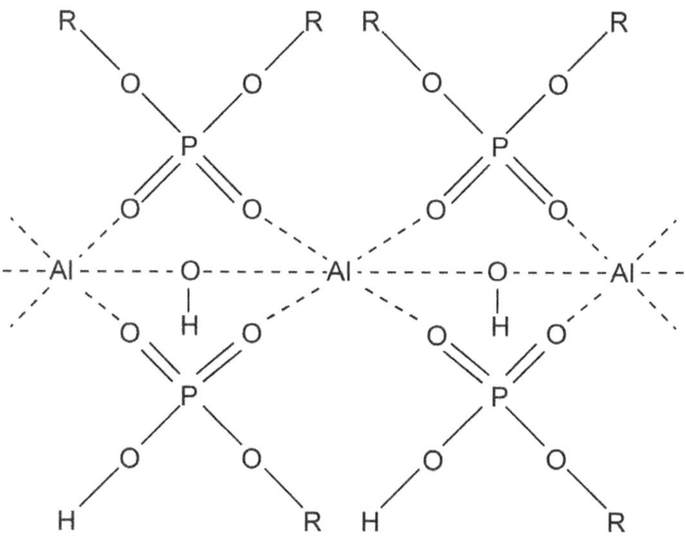

**Figure 4–2** *Probable aluminum phosphate ester bonding that forms a long-chain polymer capable of dissolving in oil to create an oil-based gel. The R-groups are 12-18 carbon chains that create the solubility in oil.*

Oil-based frac-fluids may still be used to fracture highly water-sensitive formations. Some formations contain clays that react with low-salinity water-forming hydrates that increase their volume causing them to expand within the rock pores. Some clay particles also break loose when contacted with fresh water and lodge in pore throats blocking the flow of reservoir fluids. These two actions can drastically reduce the permeability of the rock matrix (Tiab and Donaldson, 2004, 2012; Donaldson and Alam, 2008). An obvious solution is to add salts to the aqueous frac-fluid to increase the salinity and match the salinity of the formation water; however, that is not always practical. But, solutions of methanol and isopropanol suppress the hydration of clay and thus find applications where water sensitivity is a problem.

Another type of formation damage is a function of the capillary-pressure/fluid-saturation relationships in gas reservoirs that have little or no water or oil saturation. The high capillary pressures (>100 psi) that occur in the fine capillaries of shale support high irreducible saturations of water and/or oil; as much as 40% of the total porosity. Considering just water, if the shale has irreducible water saturation ($S_{wi}$) equal to 40%, but only contains 10% pore volume of water, then 90% of the pores are available for movement of gas. If water is artificially introduced (from frac-fluid) it will immediately be imbibed by capillary action until an immobile, irreducible ($S_{wi}$) equal to 40% of the pores space is attained; thus, the amount of pore space available for flow of gas is reduced from 90% to 60% resulting in a large decrease of matrix permeability. The same thing is true for shale that is under-saturated with oil; if oil is introduced from frac-fluid it will increase the irreducible saturation of oil by imbibition until the natural residual oil saturation is attained and thus diminish the permeability of the matrix.

The remedial action is to inject an alcohol to dissolve the accidentally added fluid and remove it by pumping the solution out. This procedure, though, is expensive, time consuming, and sometimes marginally effective. Laboratory analyses of cores for saturation and fluid flow properties, as well as mechanical stress properties, are therefore recommenced (Donaldson and Alam, 2008–Section 1.12).

## 4.3 Water-Based Frac-Fluids

Water-based frac-fluids are used in perhaps two-thirds of all fracture treatments, the balance is oil-, alcohol-, and foam-based frac-fluids. The most convenient source of water is often the produced formation brine. Fresh water cannot be used to fracture sedimentary rocks because they contain variable amounts of water-sensitive clays. Therefore, the fresh water is converted to a synthetic brine by the addition of potassium-, calcium-, ammonium-, or sodium-chloride. Potassium ions form the strongest bond with the negative sites in the clay molecule, preventing hydrogen bonding with water. Hence, it is used for highly water-sensitive formations. The other salts are normally selected on the basis of regional cost for moderate to low water-sensitive formations.

Water has many attributes that make it more desirable than oil-based fluids for use as a frac-fluid. It is more economical and readily available because a greater variety of chemical compounds are more soluble in water than in oils, making water easier to modify in order to meet the great variety of subsurface reservoir conditions. Water is much safer to handle than flammable oil mixtures in that it can be used for a dry gas-producing reservoir where an oil-based fluid would establish residual oil saturation in the matrix and sharply reduce the relative permeability of gas. Water easily forms high-viscosity gels, emulsions, and foams that make water more versatile than oil-based fluids.

The viscosity of water is 1.0 *cP* at 68°F; therefore, it must be thickened for use as an effective frac-fluid because suspension and transport of dense proppants requires a viscosity of 100 times or more. Guar gum (extracted from the seeds of the cyanosis-psoralioides plant) is soluble in water and effectively increases its viscosity. Chemically it is a long-chain polymer composed of *D*-mannose and *D*-galactose sugars, Figure 4–3a and Figure 4–3b. When powered guar gum is added to water, the gum hydrates (adsorbs and loosely bonds with water) forming a fluid, or gel, with a high apparent viscosity depending on concentration and temperature. The apparent viscosity of aqueous solutions of guar ranges from 10 *cP* to 100 *cP* at 80°F for concentrations of 20–80 lbs/1000 gal. Powdered guar retains 5–10% of plant materials that are insoluble in water and cause formation damage by plugging the rock matrix pores. Consequently, chemical derivatives of guar were developed to overcome this deficiency. Propylene oxide reacts readily with the hydroxyl groups on guar, producing a high molecular weight polymer [hydroxyl-propyl-guar (*HPG*)] that has excellent fracture propagation, proppant-carrying ability, and temperature stability. Other polymers used to prepare linear-polymer gels are: hydroxyl-cellulose-guar (*HCE*), carboxyl-methyl-hydroxy-propyl-guar (*CMHPG*), carboxyl-ethyl-cellulose (*CEC*), and xanthan gum. A linear-polymer gel may be prepared from a mixture of fresh water with 1–2% potassium chloride (or formation brine) and 2–5% *HPG* or *HEC* polymer. Table 4–1 lists the apparent viscosity of the polymers at two temperatures (Lyons and Plisga, 2005).

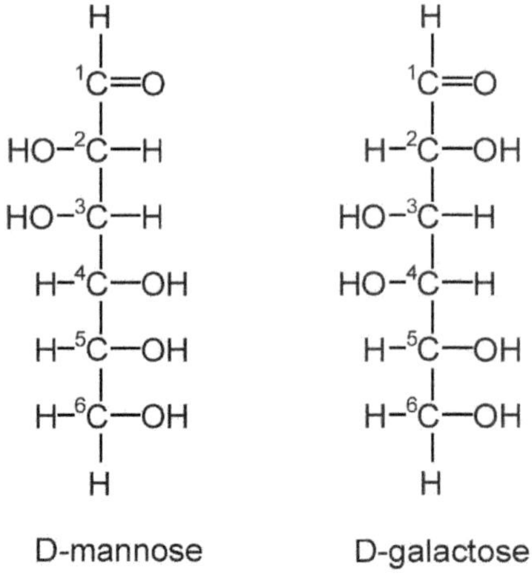


**Figure 4–3a** *D-mannose and D-galactose shown as mono-saccharides. Many sugars are related, differing only by the stereochemistry at one or more of the carbon atoms. D-mannose differs from D-galactose at carbon atoms 2 and 4 where the positions of hydroxyl groups are switched.*

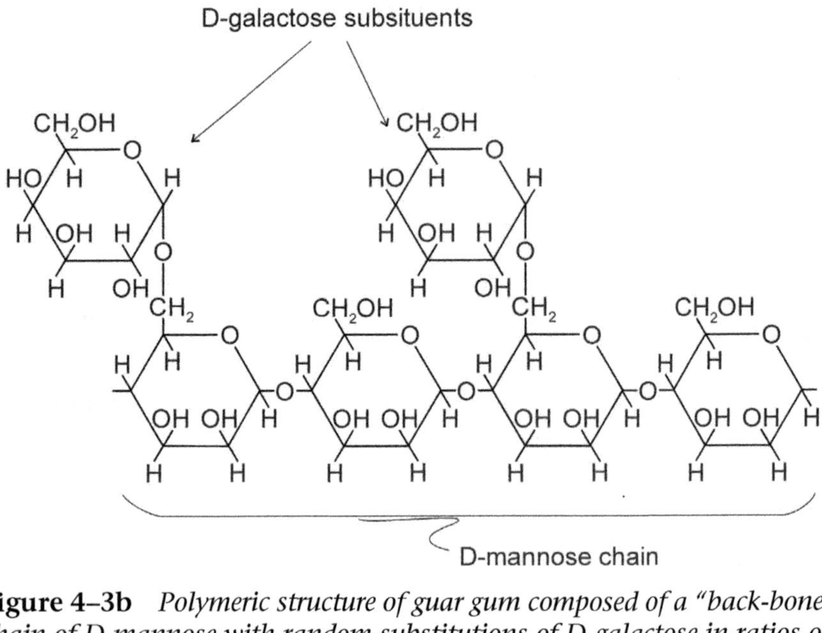


**Figure 4–3b** *Polymeric structure of guar gum composed of a "back-bone" chain of D-mannose with random substitutions of D-galactose in ratios of 1/6 to 1/8.*

**Table 4–1** Approximate Apparent Viscosities of Solution of Linear Polymers in Water (HPG = hydroxyl-propyl-guar; CMHPG = carboxy-methyl-hydroxy-propyl-guar; HEC = hydroxyl-elthyl-cellulose)

| Linear Polymer | Concentration Lbs/1000 gal | cP at 80°F | cP at 150°F |
|---|---|---|---|
| Guar | 20 | 10 | 5 |
| " | 80 | 100 | 70 |
| HPG | 20 | 15 | 10 |
| " | 80 | 110 | 90 |
| CMHPG | 20 | 10 | 5 |
| " | 80 | 90 | 60 |
| HEC | 20 | 10 | 5 |
| " | 80 | 90 | 60 |

Water-external emulsions have high viscosity and low fluid loss. In addition, they usually do not require other additives. The emulsions are generally prepared with two-thirds oil and one-third water containing 1–2% guar, *HPG* or *HEC* for increased viscosity and temperature stability ranging to 200°F.

A major advance in fracturing technology occurred in the 1960s with the discovery that polymers could easily be cross-linked, increasing the apparent viscosity and temperature stability of the frac-fluids, Figure 4–4. Guar, or *HPG,* is cross-linked with boron and the transition metals: aluminum, antimony, boron, chromium, copper, etc., forming metal chelates with the hydroxyl groups of the polymers. The molecular weight (and hence the density) of the base-polymer is increased by cross-linking, and both viscosity and temperature tolerance are increased simultaneously.

Carboyl-methyl-hydroxyl-propyl-guar is less expensive than *HPG* and is generally applicable to wells that have high bottom-hole temperatures. The biopolymer, xanthan, produced by the microbe *Xanthomonus compestris*), can be used in formations where the temperature is 300°F or less, cross-linking the polymer with zirconium and/or titanium. The temperature stability can be further enhanced to around 400°F.

The polymers used for thickening agents also act as friction reducers when used with low-viscosity fluids, which are pumped at

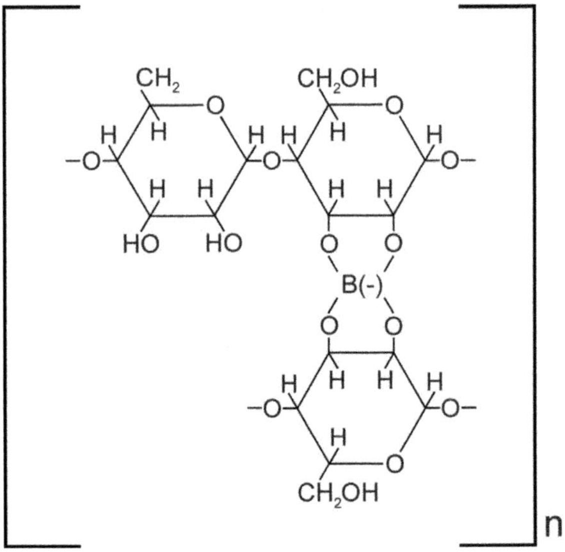


**Figure 4–4** *Idealized cross-linkage of guar (HPG or HEC) with borate salts to produce a cross-linked gel when dissolved in water. High pH (9 to 10) is required to maintain the cross-link bond.*

high rates to initiate fractures. Rapid rates of fluid flow in the well tubing promote considerable turbulence that results in loss of fluid pressure due to friction in the pipes. A large decrease of friction occurs when a small amount of polymer (10 lbs/1000 gal) is added to the frac-fluid. The polymer dramatically decreases friction loss by reducing the mobility of water molecules and thus eliminating much of the turbulent disorder of the fluid moving down the pipe.

Ferrous and ferric salts in some formations present problems from reaction with frac-mixture compounds that produce precipitates within the reservoir matrix, effectively blocking fluid flow. This is a common problem in oil fields, and experience has shown that the problem can be controlled by addition of a citric acid/acetic acid mixture to sequester the iron salts. Another widely-used compound, generally known by its acronym *EDT*, because of its general use for water treatment, is ethylene-diamine-tetra-acetic acid.

## 4.4 Alcohol-based Frac-fluids

Alcohol-based frac-fluids are usually used to fracture low permeability, dry gas reservoirs. Water- and oil-based frac-fluids will leave high residual water or oil saturations in the vicinity of the matrix near the wall of the fracture. These high saturations impede the flow of gas from the reservoir into the fractures and are difficult to remove. The alcohols (methyl and isopropyl), however, will quickly evaporate into the flowing gas stream when the fractured well is placed on production. In fact, the alcohols are frequently used for removal of "water-blocks" (high water saturations in the vicinity of the well-bore).

A mixture of alcohol, carbon dioxide and liquid-produced gas is used in low permeability, low pressure gas wells. The alcohol dissolves in the formation fluids and the gases mix with the formation gas as it is produced. This is used for very special conditions because there is considerable cost involved with the storage, mixing, pumping, etc. of the components and mixtures.

Guar gum and hydroxy-propyl-guar are used to increase the viscosity of alcohols. Guar gum will produce gels of aqueous solutions containing less than 25% alcohol, but at greater alcohol concentrations the polymer becomes insoluble. Solutions of either alcohol containing a maximum of 60% alcohol can be thickened, or gelled, with HPG. Alcohol-based frac-fluids are used sparingly for very specific conditions of low permeability dry gas reservoirs because of their high cost, danger of workers inhaling the fumes, and the requirement of high concentrations of viscosity-breakers for post-fracture clean-up of the fracture zone. The alcohols have several advantages, however, when added to acid solutions used for fracturing and etching carbonate reservoirs: they are miscible with water and remove water-blocks that tend to be established whenever water-based frac-fluids are used, and they can be used in water-sensitive formations.

## 4.5 Acid Frac-Fluid

Acid frac-fluid, consisting of 3–15% hydrochloric acid, is used when the formation material (dolomite, limestone) can be dissolved by the acid. The walls of the fracture are randomly etched by the acid forming channels that remain in place when the fracture closes after completion of the treatment. The viscosity is moderated by addition of the biopolymer xanthan or synthetic polymers.

## 4.6 Foams

Foams may be used to fracture low-permeability gas zones, shallow (low pressure) gas formations and water-sensitive formations. Foams can be prepared using alcohols, light oils, and water as the base fluid (Appendix B). A surfactant is required to lower the gas-liquid interface and stabilize the bubbles. The addition of guar, *HPG,* or xanthan polymer increases the viscosity and stability of the foam and thus its proppant-carrying ability. Without the polymer, the stability of the foam may be too short in duration for useful application. Stability, measured as the time required for half of the foam to break down (its half-life), is as little as 3–4 minutes, whereas addition of a polymer can extend the half-life to 3–4 minutes. In the laboratory, the half-life is used as an indicator of the foam quality (percent gas to foam volume). A foam quality between about 50% to 90% is necessary for a stable foam to form. Below 50% the foam generally will not form, and above 90% the foam will change to a mist (Gaydos and Harris, 1980; Holcomb et al., 1980).

Foams are prepared by introducing either nitrogen or carbon dioxide into the liquid mixture. Nitrogen is considerably less dense than $CO_2$ and does not dissolve in the liquid and hence forms a true foam. Carbon dioxide, however, is soluble in the liquid and thus more is required to saturate the liquid before foam formation begins. The dissolved $CO_2$ reduces the pH of the frac-fluid-foam to about 4, forming an acid solution that inhibits clay swelling in water-sensitive formations. Because of the low foam density the greatest amount of sand proppant that can be carried is about 6–8 lbs/gal of liquid.

## 4.7 "Slick-Water"

Slick-water is fresh or saline water containing high molecular weight polyacrylamide polymers (0.01 to 0.1%) that inhibit turbulence and thus reduce pressure loss due to the friction of turbulent fluids as they are pumped to the formation. Slick-water also will contain a biocide to avoid aerobic bacteria slime and control *desulfo vibrio* that metabolizes sulfates and organic-sulfur compounds by reduction to hydrogen sulfide that can sour an entire reservoir. Methanol, naphthalene, quaternary-amine, or chlorine-dioxide is used as a biocide in concentrations of 0.005 to 0.1%. Surfactants such as ethyl-monobuyl-ester are used to enhance proppant carrying ability a scale inhibitor (hydrochloric acid, ethylene-glycol, phosphate-esters, or phosphate-polymers) and surfactants to increase the proppant carrying properties (butanol, ethylene-monobutyl-ester).

The reduction of friction allows very high pumping rates (100 bbl/min) and use of large volumes of frac-fluid (one to five million gallons per fracture treatment). "Slick-water" is most effective in hard, brittle, shale; and its success as a frac-fluid is attributed to enlargement of natural micro-fractures in the shale bed. The sand-carrying capacity of slick water is very low, but is not an important factor to the success of the fracture treatment where the shale is brittle. "Dune" development of the proppant (deposit near the well-bore with carryover of grains of sand as high-velocity fluid is pumped into the fracture system) carries enough of the proppant far into the fracture system so that the fractures do not lose much of their width when the well is placed on production. Loss of frac-fluid by true leak-off into the matrix of the shale is low. It only occurs in the extended fracture network and opened bedding laminations.

"Slick-water" is most useful for deep brittle shales (gelled fluid is used in ductile shales, high permeability rocks, and where large amounts of proppant must be carried into the fractures). If the shale is water sensitive, salts (especially potassium chloride) can be added up to about 3%. Any higher concentrations result in "salting-out" of the polymers and other additives (Daneshy, 2011; Gupta, 2009; King, 2010; Sun, et. al, 2011).

## 4.8 Surfactants

Hydrocarbon-based surfactants (carbon atom "tails" composed of carbon-hydrogen bonds and a carbonyl group...soaps) are used extensively to stabilize oil-in-water emulsions, Appendix B.

Flourocarbon surfactants (where fluorine atoms replace the hydrogen atoms in the carbon-chain) are used for sandstone formations containing clay particles. These clay particles improve the back-flow of aqueous solutions from the matrix during the back-flow phase of the fracture treatment, leading to establishment of a water-block to production of the hydrocarbon reserves.

The fluorocarbon surfactants yield considerably lower interfacial tension and water-rock contact angles than do the hydrocarbon surfactants. Thus, they can be effectively employed at much lower concentrations (Donaldson and Alam, 2010). The capillary pressure equation explains the phenomena: the capillary pressure between any two phases is a direct function of the surface tension and the contact angle, Eq. (4.1):

$$P_c = \frac{2\sigma_{ift} \bullet Cos\theta}{r} \tag{4.1}$$

The fluorocarbon surfactants strongly adsorb onto the rock surface and cause a decrease of surface tension ($\sigma_{ift}$) and an increase of the water-rock contact angle ($\theta$). The result is a decrease of capillary pressure that enables mobility of the reservoir matrix brine and frac-fluid, causing fractures to develop within the matrix pore network.

## 4.9 Clay Stabilizers

The hydration of clays brought about by injection of low-salinity water and acid solutions is controlled by addition of ammonium chloride. Formation brine is preferred because it is compatible with the formation fluid and does not promote clay swelling.

There are many non-swelling clays in formations, but they are usually prone to migrate in sandstone reservoir and lodge at narrow pore throats, causing considerable reduction of permeability. Especially troublesome are chlorite, illite, and kaolinite clays. However, zirconium chloride forms a hydrate polymer in water that adsorbs on the clays and binds the particles to sand-grain surfaces (Veley, 1969). This stabilizer is used in a pre-flush before injection of the frac-fluid rather than as a direct additive to the frac-fluid mixture.

## 4.10 Temperature Stabilizers

The acrylamine and cellulose-derived polymers are the most temperature-stable of the polymers used to thicken frac-fluids. However, all of the polymers are sensitive to pH because the hydrolytic bonding is degraded by acidic conditions, especially at high bottom-hole temperatures. Therefore, the pH of frac-fluids is normally maintained at values between 8 and 10 (basic solutions) to enhance stability by elimination of the presence of hydrogen ions.

Entrained oxygen (from storage and mixing operations at the surface) is also detrimental to the polymers as the temperature of the frac-fluid is elevated. Methyl alcohol and sodium thiosulfate are effective oxygen stabilizers and are used for fracturing deep formations.

## 4.11 Fluid-Loss Additives

Fracturing fluid loss to the formation matrix and micro- and macro-fractures occur as soon as the fractures are formed. This loss of fluid is very detrimental because it impedes further extension of the fractures. Therefore, additives with high viscosities and particles (which build up on the face of the fracture like a filter-cake as it forms) are used to stem the loss of frac-fluid. Effective control of fluid loss is obtained from particles within a size range that will bridge the pore openings combined with a long chain polymer to act as a plastering agent and form a filter-cake on the walls of the fracture. A typical sandstone will have a pore size range from 1.0 to 25 μm. An effective particle size range is 1 to 75 μm.

Perhaps the best fluid-loss additive that was developed for oil-based frac-fluids is composed of calcium carbonate particles coated with an oil-soluble surfactant. However, it is incompatible with the aluminum-ester (napalm) gels where oil-soluble resins and non-oil-soluble agents such as silica flour, salt, or benzoic acid are used.

Silica "flour" is commonly used with water-based frac-fluids. It is ground silica that will pass through a 325-mesh screen (43 μm). Non-swelling clays, talcs, and guar gum also are used, all of which form filter-cake layers on the fracture walls with very little penetration into the rock pores.

Fluid loss in tight sandstones (< 1 md) is best controlled by addition of about 5% diesel oil, with a surfactant, to form a micro-emulsion. This cannot be used in high-permeability reservoirs where the pore size is large enough to permit penetration of the small diameter micelle emulsion. Perhaps the most cost-effective fluid loss additives are mixtures of silica flour with guar gum as the plasticizer.

## 4.12 Viscosity Breakers

After the fracture has been initiated and the proppant properly placed in the fracture, the fluid-loss additives that formed a filter-cake on the walls of the fracture must be removed and the high viscosity of the frac-fluid must be reduced before the well can be placed on production. Fine particles in the proppants that can bridge the pore channels also have to be removed. This can only be accomplished if the viscosity of the fluid in the fracture is lowered to a "thin" (viscosity approaching that of water) fluid that can be easily pumped back out of the fracture followed by the reservoir fluids (oil and/or gas and brine). All of the viscosity breakers degrade the polymers used in water-based frac-fluid. They are added to the mixture just before injection begins or pumped into the fracture and allowed to rest for 5–24 hours (depending on the type of frac-fluid) before production is resumed, which begins by pumping out the frac-fluid mixture ahead of the formation fluids. As they are pumped into the well hole, acids and bases will react with aluminum-phosphate-ester gels if the bottom-hole temperature is less than 100°F. Above 100°F, a viscosity breaker is not required for aluminum-phosphate-ester gels because they break naturally at higher temperatures.

## 4.13 Biocides

Biocides are required in frac-fluids as soon as they are prepared, during storage in tanks, and as they are injected into the wells. Aerobic bacteria, stimulated by entrainment of oxygen during mixing, can rapidly destroy the sugar- and cellulose-based polymers during preparation and storage, adding considerable cost to the fracture treatment. In addition, aeration during mixing also introduces spores of anaerobic bacteria, especially *desulfo vibrio* that can grow in the reservoir while metabolizing sulfur and producing hydrogen sulfide to completely sour the reservoir. About half of the biocide is added to the tanks just before the mixed frac-fluid is temporarily stored for accumulation and gel formation. The balance of the biocide is added just before injection into the well.

Some of the biocides are carbamate-gluteraldehydes (used in acid-frac-fluids) and widely used for general frac-fluid preparations (Kalish, 1964).

## 4.14 Buffers

During mixing, a basic solution of sodium carbonate or bicarbonate is used to ensure dispersion of the polymers in the solution, but hydration of the polymers (to increase viscosity and for gel formation) will not occur in basic solutions. Hence, buffers (low molecular weight organic acids such as acetic-, formic-, adipic-acids) are used to lower and control the pH for gel formation.

## 4.15 Frac-Fluid Preparation

Two methods for mixing the frac-fluids are used. Batch mixing is generally used for oil-based frac-fluids. Most of the ingredients are added to the base fluid and stirred in a single large tank (500 bbl or more capacity). The fluid-loss additives and gel breakers are metered into the gelled frac-fluid as it is pumped into the formation. The fluid-loss additives will steel-out in the tank when the mixing is completed and cannot be pumped into the reservoir in the proportions that are necessary. The gel breaking fluids must be added at the last minute because they would cause reduction of the gel viscosity if they are in contact with the gel in the tank.

On the surface, batch mixing in a tank offers better control of the frac-fluid because it can be tested to insure that it has the physical and chemical properties required to bring about the desired fracture treatment. However, it has several disadvantages: should the fracture project be delayed after the mixing for some technical or management reason, the gel may have time to degrade chemically or by action from bacteria, causing the entire mixture to be discarded. The entire contents of the mixing tank are never delivered to the reservoir, 5% or more may be left in the tank at the end of the job, representing a financial loss due to the cost of materials as well as a labor loss because the remaining fluid in the tank must have the gel broken for fluid disposal and the tank must be cleaned out.

Some of the ingredients of water-based fluids may be pre-mixed in a tank such as the completely soluble polymer, salts for clay stabilization, and the bactericide. It is best for the polymer to have some time to hydrate completely in the tank before it is injected. Cross-linking compounds for guar and its derivatives (titanium and zirconium salts) are best metered into the frac-fluid as they are pumped into the reservoir. In consort with the cross-linking compounds, the pH must be adjusted to promote cross-linking to take place. For example, borate cross-linking occurs at a pH of 10 or higher. Hence the pH is increased using sodium hydroxide metered into the frac-fluid as it is pumped into the reservoir.

Continuous mixing (metering in the necessary additives through a multiport manifold as it is pumped into the well) does not have the disadvantages mentioned for batch mixing. However, considerable experience in monitoring, controlling, and timing of the metered additives is necessary since sampling to determine if all of the materials are properly blended during the pumping stages is very difficult to achieve. The frac-fluid must meet a number of specifications (as discussed previously) to be effective.

## 4.16 Conclusion

This brief discussion of frac-fluids is intended to reveal the most prominent components that go into making a composite fluid to bring about a desirable pattern of fractures in a specific gas-shale reservoir. The additives are, without doubt, changing as newer chemical compositions (many of them proprietary) are developed to address the variability of gas-shale reservoirs. The implementation of hydraulic fracturing is absolutely necessary for the production of this newly-discovered energy resource, which is in many ways vital to modern civilization.

CHAPTER 5

# Field Implementation of Hydraulic Fracturing

## 5.1 Introduction

Methods of hydraulic fracturing are evolving at a rapid pace, attempting to meet the requirements of complex geology, highly impermeable production zones, and environmental concerns. Though fracturing has been conducted for over 50 years and the basic principles of fracturing have not changed substantially over the last 20 years, our understanding of the fracturing process has improved tremendously while its implementation has become much more effective. Fracturing is conducted more frequently now than ever before. It is estimated that more than 90% of the wells are now fractured. Such a high rate of fracturing is the result of re-entry into former "depleted" reservoirs and tight (low permeability) formations with horizontal wells. Reservoirs that were considered to be depleted because production from vertical wells was reduced to a trickle, in reality still contain 50% to 60% of the original oil that was in place in the reservoir which is trapped by the capillary pressure properties of the rock matrix. Horizontal wells and hydraulic fracturing produced a new era of oil and gas production from the so-called depleted reservoirs (Halliburton, 2012).

Hydraulic fracturing operations are overseen by operators and service companies to evaluate and document, in as much detail as possible, all of the events of the treatment process. Every aspect of the fracturing operation, from pressure measurements and pumping rates to volumes of additives and environmental protection, is monitored, recorded, and retained for future reference. It is not unusual to have between 30 to 40 people monitoring the entire process of a fracture treatment. Horizontal wells can be in excess of 5,000 feet long and are generally fractured in stages, allowing the operator sufficient time to make any changes necessary to address site-specific variations such as: formation thickness, the presence and absence of natural fractures, and the proximity to other fractures. Nevertheless, time at the field site is very limited; therefore, the fracturing crew must have any and all engineering designs and backup plans ready for implementation when they arrive at a site.

## 5.2 Protecting Groundwater

Surface water (lakes and streams) and subsurface groundwater (often a fresh water aquifer) is a vital, natural resource that is used for sustaining lives. In many places, groundwater is the principal source of drinking water, and in some places, the sole source aquifer. Protection of groundwater is critical to sustain developments in the production of oil and gas.

The hydraulic fracturing process injects fluids into the reservoir rock matrix through a pipe in a well that is isolated from groundwater through several layers of steel pipes called casings, Figure 5–1. These casings are designed and cemented to the ground or to one another in a fashion that neither fracture fluids nor flow-back fluids flowing through the inner tube can enter a groundwater aquifer. The casings are concentric steel pipes that provide multi-layers of protection to the groundwater. The outer casing, called *conductor casing,* is set several hundred feet deep from the surface and is used to guide the drilling operation. The *surface casing* is set inside of the conductor casing and to a depth below the fresh water aquifer. It is then cemented in place by pumping cement down the casing, forcing it to return to the surface in the annulus between the casing and the formation and the casing and conductor well. Thus an impermeable cement seal is placed between the casing and the ground formation that holds the casing in place and forms a double barrier to leakage of fluid from the surface casing into the formation. The *production casing* is the inner casing that is placed all the way down to the bottom of the well, thus separating the producing zone from the other formation layers. Other intermediate casing, or tubing, may be used within the production casing to isolate other zones of interest. Frequently, a working pipe is placed within the production pipe for injection of the frac-fluid and conduction of the flow-back fluid when it is produced at the end of the fracture treatment (refer to Figure 2–3). With such redundancy of casings and cementations, the possibility of frac-fluid or flow-back fluid leaking into the groundwater is very remote. It is, however, important that periodic inspections are conducted to ensure

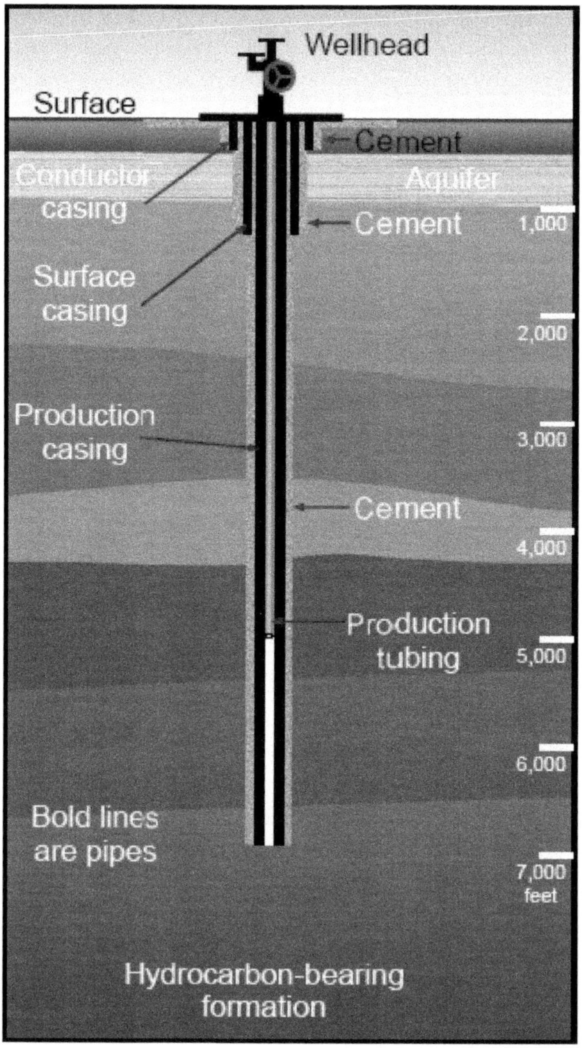


**Figure 5–1** *Construction of a typical well. (Source: EPA Hydraulic Fracturing Study Plan, 2011)*

the integrity of the casing and cementation in case corrective action is required.

Operators perform a variety of checks to confirm that the desired isolation of each zone is intact, ensuring that the casing used has sufficient strength and that the cement has properly bonded to the casing. An acoustic cement bond log is run with computer imaging to verify the mechanical integrity of casings. State oil and gas regulatory agencies often specify the required depth for protective casings and regulate the time that is required for cement to set before the continuation of deep drilling. These requirements are typically based on regional conditions.

Generally, state regulatory requirements for protecting the groundwater are quite stringent. Even when not required by regulations, operators can conduct a baseline survey of the location and quality of groundwater before beginning any activities, and in some cases they may install one or more monitoring wells in close proximity from the production or injection well to safeguard against contamination by inadvertent leakage of fracturing or production fluid into the groundwater aquifer. In most cases, monitoring wells are placed on the down-dip side, but when there is the possibility that low-density hydrocarbons might migrate up-dip a monitoring well will be placed on the up-dip side as well.

Groundwater is generally only a few hundred feet from the ground surface whereas the hydrocarbon formations that are fractured occur at thousands, and sometimes tens of thousands, of feet below ground under many layers of impermeable rock formations called seals, or confining layers, between the hydrocarbon containing formation and groundwater. The geologic seals are the reason why hydrocarbons are trapped in certain locations and cannot move. Therefore, migration of fracturing fluid or flow-back of hydrocarbon fluids through natural fractures all the way up to the shallow groundwater aquifers is very unlikely. However, in the remote case where a well is located close to a natural fault whose closure is compromised such that there are some channels that can conduct fluids vertically into a communicating aquifer, some contamination of the aquifer is possible. However, this is a very remote possibility because the initial surface seismic survey, which is used to characterize the gas-shale formation, would find the fault line and no reputable company would conduct operations near enough to involve the fault system.

Groundwater contamination by fracturing or flow-back fluid would not occur though the well casing because of the many protective layers of casing that are used. Fractures extending from the reservoir rock to groundwater are unheard of because the intervening rock presents multiple seals to vertical flow of fluids. Contamination is therefore limited to spills or leakage from equipment on the surface. The surface facilities may be relatively close, only a few feet above the groundwater in some cases, and often with little or no natural barriers to prevent seepage into the groundwater. Thus the surface facilities frequently have adequate redundancy for protection against spills. In any case, should a spill occur, a *Spill Prevention Control and Counter-*

*measure* plan should always be in place for a step-wise guide for protection and cleanup. The spill must not be allowed to reach groundwater or surface water. Oily components of the fracture and other fluids can damage the groundwater severely, and it is very difficult to remediate. Pump-and-treat methods used for treating groundwater are not very effective where hydrocarbons are concerned. In addition, the small percentage of chemicals used in the fracture fluids could render the groundwater unsuitable for human consumption and use for years before it is finally dissipated by migration and dilution.

## 5.3 Waste Water Management in Hydraulic Fracture

Wastewater generated from fracturing operations is frequently disposed of by injecting the flow-back fluid into deep formations that are geologically isolated and have no significant impact on the environment. These disposal injection wells are classified by the US EPA as Class II injection wells and have served the petroleum industry for disposal of oilfield brines and the manufacturing industry for disposal of liquid wastes that cannot be treated by currently known surface treatment methods (Figure 2–3 illustrates the detailed construction of Class II disposal wells). However, the great volumes of fracturing flow-back water that are encountered from some extensive fracture treatment programs (where multiple lateral horizontal wells are fractured) is sometimes greater than a single disposal well can handle. For example, it is estimated that the Eagle Ford Shale required over six million gallons for fracture fluid per well, and more than 20% flowed back and had to be disposed of at the surface. The rate and total volume of fluids that an injection disposal well can handle is restricted by the permeability, porosity, and reservoir fluid pressure. Currently, there are about 30,000 Class II disposal wells in the United States, but new site specific wells will have to be constructed with very demanding specifications to handle the amount of waste generated from gas-shale fracture treatments. Hence, other methods of disposal, or recycling, of flow-back fluids are aggressively under research investigation because they will need to be developed to ensure safe handling of the wastewater generated from fracturing operations.

The geological evaluations of reservoirs at the site of fracturing operations are very important not only for characterizing of the gas-shale formation but also for geological structures that can support an injection well if one is needed. Any fault zones detected in the vicinity by the seismic survey must be considered if an injection well is planned because high pressure injection of liquid waste near a fault zone can lubricate joints that are under tectonic stress, causing slippage that can produce minor local earth-tremors that raise issues with a local population. Also transporting waste fluid by trucks from a fracturing site to a disposal well through highly populated zones can cause traffic congestions and create safety hazards. Transporting a million gallons of fluid can require 200 or more trips by trucks. This may be avoided by transporting the waste fluid through pipe lines but these are generally not available, too expensive, time consuming, and difficult to construct.

The volume of water going to the disposal well can be reduced by treating and recycling the water. The challenge with treatment of flow-back water for reuse for subsequent fracturing is more logistical in nature than technical. The volume of water that must be treated in a short period of time of weeks from each well can be a million gallons or more. The volume can range from 20% to more than 70% of the original fracture fluid volume. In some cases, flow-back of fracturing fluid in produced water can continue for several months after gas production has begun, requiring the treatment unit to be available for the period. The treated water would have to be transferred and stored in a new location for fracturing in a remote location. The treatment process itself may require special techniques due to the high salinity of some flow-back from fractured wells. Reverse osmosis might be used, but in many cases it may not be applicable. Therefore, considering these constraints, some novel wastewater treatment methods for recycling are being developed as discussed below:

- *Distillation*—Fountain Quail Water Management of Fort Worth, Texas, has mobile and stationary recycling units that distill the waste water by applying heat to separate high concentrations of salt from the water. The concentrated salt water produced is disposed in disposal wells while the treated water is recycled for subsequent fracture jobs. Using this process, instead of hauling the flow-back fluids to a disposal well, it is stored in tanks on location and distilled with a mobile unit or piped to a treatment plant. Natural gas produced on location is used to operate the distillation columns.
- *Publically owned Municipal Treatment Facility*—This is another method of reducing disposal by injection wells is treatment by a nearby publically-owned treatment facility which is currently being used by the Barnett Shale Water Conservation Company and Brazos Bend Energy Services. Although this type of treatment does not allow the recycling of the water for additional fracture treatments, it does allow the water to

remain in the hydrologic cycle that can help the recharge of an aquifer. Regulatory permission, proximity to a publically owned facility, and its ability to handle the additional short term load are some of the constraints of this method of recycling.

- *FilterSure*—This is a treatment process uses electro-coagulation followed by filtration to treat flow-back water. The system is still in the developmental phase and is currently targeted for Marcellus Shale.
- *CleanWave*—This is a process that also uses electro-coagulation for treating flow-back water. The process is designed to recycle brines, reducing the cost of salt addition to fracturing fluids to control swelling of formation clay that can severely damage the permeability of the reservoir.

## 5.4 Fresh Water Management in Hydraulic Fracturing

Hydraulic fracturing consumes huge amounts of fresh water that can have an adverse effect on fresh water in arid regions. Two major shale formations in Texas that have high hydrocarbon potential are the Barnett and Eagle Ford shales. It is expected that thousands of wells will be drilled in these formations, requiring billions of gallons of fresh water for the fracture treatments. Pumping large quantities of water (1–6 million gallons per well) in a short period of time (a few weeks) for hydraulic fracturing will have a detrimental effect on the fresh water resources in the vicinity. Recharge of fresh water aquifers in the area is normally very low, requiring years to do so. Unlike farming (which also consumes large quantities of fresh water) where a major portion of the fresh water used for irrigation trickles back into the aquifer as recharge, the major portion of water used for hydraulic fracturing is lost in the formation in which it is injected. The withdrawal from fresh water aquifers and loss of such huge quantities of fresh water can impact the availability of an underground source of drinking water. It can also cause subsidence where the porosity of the aquifer collapses and is permanently lost, but this depends on a combination of geologic conditions as well as the withdrawal volumes (Chilingarian et al., 1995). In coastal areas, large withdrawals of groundwater can cause salt water intrusion. It is important that water withdrawal be closely managed in order to sustain fracture treatment activities. Regulatory agencies have (and in some cases are in the process of developing) guidelines that are protective of the groundwater.

In Texas, the Railroad Commission has been monitoring the use of water in hydraulic fracturing (Railroad Com. of Texas, 2011). The Barnett shale, one of the most active drilling and fracturing sites in the United States, is situated in North Texas. The formation is estimated to contain over 27 trillion cubic feet of gas. Extraction of this resource (even a fraction of it) requires extensive fracture treatment for stimulation of production from the very low permeability shale matrix and will exert a tremendous demand on local fresh water sources. Increasing water use due to growing population, drought, and Barnett gas-shale development has raised concerns about water availability in North-Central Texas. In January of 2007, the Texas Water Development Board (TWDB) published a study report of a 19-county area in North Texas that includes the Barnett gas-shale development area. This report, the *"Northern Trinity/Woodbine Aquifer Groundwater Availability Model, Assessment of Groundwater Use in the Northern Trinity Aquifer Due to Urban Growth and Barnett Shale Development,"* includes estimates of water used in Barnett gas-shale development. The report shows that the Trinity and Woodbine Aquifers provide about 60% of the water used for the development of Barnett gas-shale. Although this accounts for approximately 3% of all groundwater used in the entire study area in 2005, it can constitute a much larger percentage of groundwater usage of a local area where fracturing is concentrated. Typically, groundwater provides for a greater percent of total supply of water in rural counties and a smaller proportion of total use in more urban counties. Therefore, increased groundwater use for any purpose will have a greater impact on the rural areas addressed by the study.

The TWDB estimates that the Barnett gas-shale development consumed about 2.4 billion gallons of groundwater in 2005, and the consumption has the potential of increasing to over 8 billion gallons in 2025. This corresponds to an estimated potential increase in groundwater used from 3% in 2005 to 13% in 2025.

The Eagle Ford gas-shale is located in South-West Texas and covers an area of approximately 3,000 square miles (see Figure 5–2). Due to the success in producing both oil and gas from this formation, great interest for increased drilling and fracturing to increase production has developed since 2008. The Eagle Ford gas-shale sits in an area that is arid and thus availability of fresh water is limited. The aquifer that supports the region is the Carrizo Wilcox aquifer that spans a large area across South Texas, providing drinking water for towns, such as Cotulla and many other communities in addition to ranches and farms that depend on it for agricultural use. Nevertheless, large quantities of water are now being used from the aquifer to support oil and gas production from the Eagle Ford gas-shale. To address the

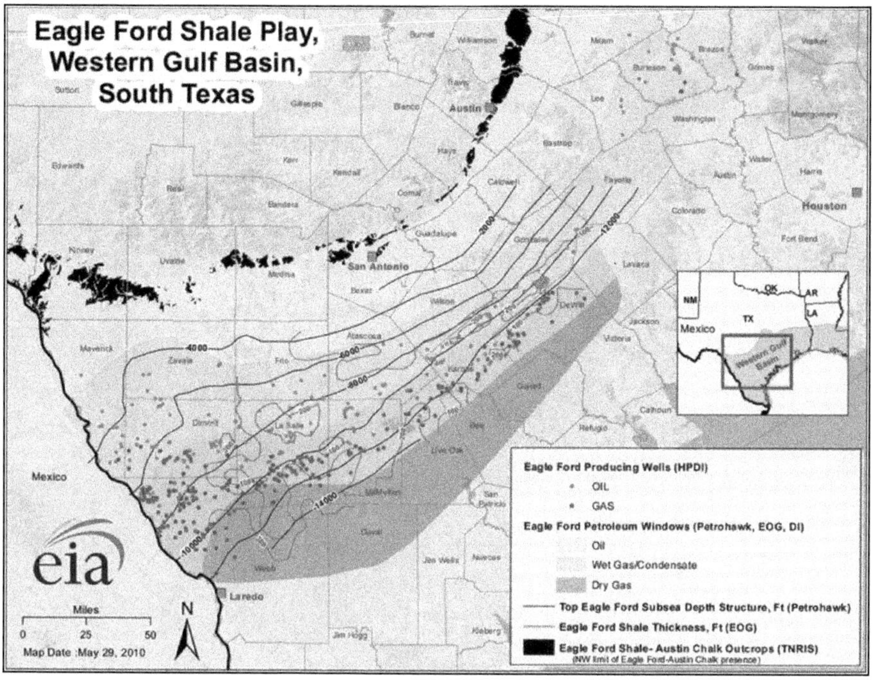


**Figure 5–2** *Map view of the Eagle Ford Shale. (EIA, 2010)*

water usage issue, water recycling is being considered, as discussed earlier.

## 5.5 Reducing Surface Disturbance

Drilling and reservoir fracture operations only last for a few weeks, but they can have a significant impact on the surface soil from the movement of heavy vehicles, installation of well-pads and surface equipment, and the storage of inventories. Reduction of the "footprint" associated with drilling and fracturing operations helps in the reduction of surface disturbance. It is estimated that to develop shale-gas in a 1-square mile section (640 acres) of land, 16 vertical wells (each located on a separate well pad) would be required because of the low drainage area of each vertical well. On the other hand, if horizontal wells are used, a single vertical well pad could support six to eight horizontal wells, extending radially, that would be capable of draining an equivalent area or even more, thus reducing the footprint for operation substantially (US Dept. of Energy, April 2009). It is estimated that on the average three to four vertical wells may be replaced with a single horizontal well, and only 7.5 acres of land will be disturbed by four horizontal wells from a single pad for development of the Fayetteville gas-shale of Arkansas. On the other hand, if vertical wells were used to access the same gas resource, 77 acres of land would be required. This ten-fold difference in development "foot-print" highlights the desire for operators to use horizontal wells.

## 5.6 Controlling Noise, Lighting, and Traffic at Fracturing Job Sites

The hydraulic fracturing operation is a continuous process that cannot usually be stopped after it is started without considerable expense. Round-the-clock work during drilling and fracturing operations is normal, and it can create noise, lighting, and traffic issues. Since several of the more productive gas-shale formations are located near populated areas, fracturing jobs are frequently conducted in such locations. When operating in these areas, care is taken to ensure protection of the environment from noise, lighting, and traffic pollution of the nearby neighborhoods. This is achieved by using sound and light protective devices and also controlling traffic to the maximum extent possible. Additionally, the technology of directional drilling allows the operators to reach productive formations from a considerable distance, especially when they are under structures or in sensitive environmental areas. An example of this is the Barnett gas-shale that is in Northern Texas near the major cities of Dallas and Fort Worth. Fracturing practices there have been modified to comply with local laws that have been implemented to minimize the impact on communities. These laws are evolving to suit the needs of specific locations, requiring that fracturing and other developmental activities must be conducted by: (1) maintaining a certain minimum distance from the population centers, (2) using a lighting system, such as directional lighting, that has a minimum effect on the surrounding areas but provides enough lighting for the safe operation for the workers, and (3) using highly effective mufflers on compressors and pumps, as well as sound barriers to minimize noise pollution affecting surrounding communities.

To minimize traffic congestion, work schedules are adjusted and/or traffic volume is reduced for transporting items (such as flowback fluid) by pipeline instead of tanker trucks. In the Barnett gas-shale area around the Dallas-Fort Worth International Airport, operators have constructed pipelines to transfer produced water from well sites to disposal facilities, thereby reducing traffic and potential damage to roads. Other ordinances: (1) control speed limits to address safety issues, (2) impose a tariff on vehicles over certain weights to collect funds for repair of roads, and (3) impose a limit on vehicle operation times in order to reduce congestion during peak traffic hours.

## 5.7 Technical Considerations for the Success of Hydraulic Fracture Treatments

Success in the art of fracturing is indicated by the increased rate of production and high ultimate recovery. Even when higher rates are observed in the field, understanding the mechanisms that cause the increased rate is essential to being able to sustain the higher rate (or even improve upon it) and to attain higher ultimate recovery from the reservoir. Productivity increase from fracturing depends on many factors such as the degree of formation damage and fracture conductivity. Hydraulic fracturing, although responsible for increasing productivity substantially, is a very expensive production stimulation process that involves a huge amount of equipment and manpower and cannot be implemented without detailed planning. Two major aspects of hydraulic fracturing that must be closely monitored are the logistical aspect and the operational aspect (Allen and Roberts, 1994).

*Logistical Aspect*—Fracturing uses a large amount of resources over a short period of time. To optimize the fracking operations, logistical aspects become very important:

- A fracturing job requires enough space around the well-site to accommodate the pumping trucks, frac-fluids, proppants, blender, chemicals, and control center. The site layout of a typical fracturing operation is shown in Figure 5–3.
- Fracturing operations can be hazardous because high pressures and chemicals are involved in the process. Therefore, manned equipment should be at a safe distance from the wellhead.
- Tank bottoms carrying fracture fluids should preferably be conical to minimize residuals left over in the tanks, and at least 5% additional fluids over the design volume should be available to adjust for tank bottom loss.
- Unobstructed space for emergency and refilling vehicles must be provided.
- Redundant equipment must be available as backup to cover any malfunction of critical equipment. For example, two pumping blenders should be rigged up side-by-side and positioned so that either can be replaced by a standby blender

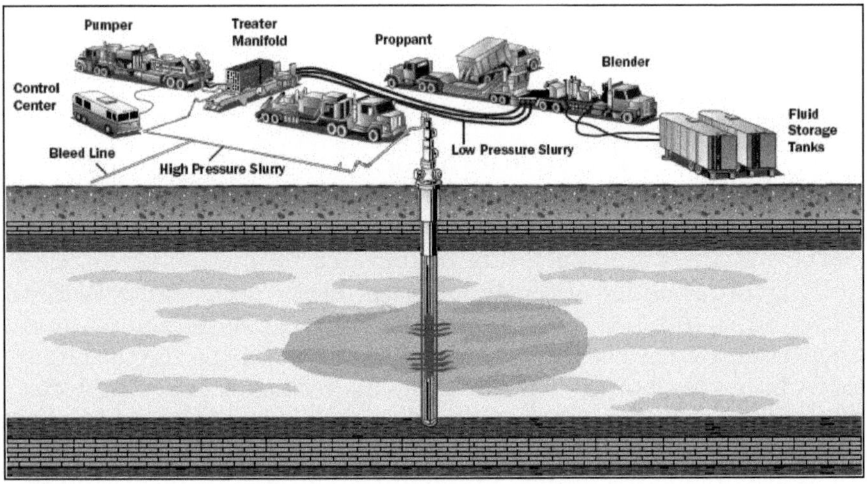


**Figure 5–3** *Site layout of a typical fracturing operation. (US Department of Energy)*

if necessary. In addition, standby high pressure pumps, viscosity thickening cross-linker pumps, and instrument vans should be ready for operation.

- Proppants must be analyzed to ensure that they meet the mesh size required according to design. Proppants that do not meet the specifications can damage the producing formation.
- Water and oil composition should be verified to ensure their compatibility with the gelling system. Produced water with oil and gas, if used instead of fresh water, may require substantial treatment.
- Before preparing the frac-fluids, all tanks must be properly cleaned to remove bacteria. Sulfate-reducing bacteria produce hydrogen sulfide and sulfite ions that can precipitate iron sulfide in the formation and clog the production channels. Bacteria also produce enzymes that can interfere with cross-linking of compounds used to increase the viscosity of the fracture fluid. The viscosity is kept high when injecting proppants and is reduced for flow-back. Enzymes can be removed by reducing the pH to less than 2 (a very acidic and corrosive condition) as long as it does not adversely impact the cross-linking process (refer to Chapter 4 for more on fracture fluids).
- All valves need to be tested for the maximum pressure.

*Operational Aspect*—Fracturing operations involve several steps that must be properly implemented and closely monitored as follows:

- It is critical to monitor the fluid and proppant rates of injection accurately because they are responsible for creating the fracture and defining the fracture length and the width. Instruments used to monitor fluid and proppant injection rates are generally quite accurate but they must be calibrated before use and additional field techniques must be used to verify the instrument's accuracy. An accurate field verification technique is measuring the volume actually removed from a fracture tank in a given time period.
- Fluctuation of the treating pressure can damage the wellbore. Fluctuations can be caused by: (1) instrument malfunction, (2) changes in gel properties, (3) variations in proppant concentrations, and (4) problems with cross-linked fluids. If the viscosity of the fluids deviates from the design viscosity, it will not deliver the correct amount of proppants to the fracture openings. This will be reflected by pressure deviation and must be corrected immediately. When slick-water is used for the frac-fluid, flow rate variations are reflected by pressure variations and must be verified and corrected to meet the design requirements.

## 5.8 Case Studies of Hydraulic Fracturing

### 5.8.1 Assessing Fracture Performance Using Quantitative Data (the RH Field)

Dmour and Shokir (2008) used methods to analyze parameters of hydraulic fracturing operation in a field (the RH gas field) that can help optimize hydraulic fracturing process and forecasting of well performance. Methods of analysis usually rely on pressure transient and production data.

The RH gas reservoir, consisting of clean sandstone and siltstone, is located in the eastern part of the panhandle of Jordan and was discovered in 1986. The field produces mostly dry gas with small amounts of water from a tight sandstone matrix that has 7 to 15% porosity and less than 0.1 md permeability and contains natural fractures. The reservoir properties obtained from the RH-x well are shown in Table 5–1. The data were obtained from log and core analysis and static pressure surveys. The fracture gradient was calculated from the fracture treatment data.

**Table 5–1** Reservoir Properties of RH-x Well

| Formation Type | Depth (ft) | Reservoir Pressure (psi) | Reservoir Temperature (°F) | Net Gas Pay (ft) |
|---|---|---|---|---|
| Sandstone | 8,937 | 3,400 | 300 | 29.5–42.6 |

| Formation Type | Gas Porosity (%) | Fracture Gradient (psi/ft) | Water Saturation (%) |
|---|---|---|---|
| Sandstone | 7–15 | 0.82 | 35 |

The RH-x well is located in a very heterogeneous low permeability sandstone reservoir with three multilayered producing zones. Assessment of the effectiveness of the hydraulic fracturing was conducted by:

- Modified Isochronal Test
- Pressure transient analysis
- Production Data Analysis

***Modified Isochronal Test (MIT)*** is performed to determine the capacity of the well to flow and is called the deliverability of the well. An MIT was conducted on the RH-x well and the reservoir parameters determined were used to design the fracture job. After the MIT, the well was stimulated by hydraulic fracturing followed by a period of pressure build-up and a post-fracture MIT test. Results of the MIT are presented in Tables 5–2 and 5–3.

The results show that the flow rate of the extended flow after fracturing increased from about 12 million standard cubic feet per day (12 MMscfd) at 1,545 psi bottom-hole pressure (BHP) to about 23 MMscfd at 2,083 psi BHP.

***Pressure transient analysis*** is the analysis of pressure changes over time and has been used successfully to obtain estimates of reservoir and fracture properties. Bottom-hole pressure data was interpreted using a computer program to estimate reservoir parameters and to define the reservoir model with semi-log and log-log plots and type-curve matching. In general, the pressure response in an ideal hydraulic fractured well is expected to behave as follows:

> *The wellbore storage effect is observed at the beginning of the pressure transient analysis and is usually large for a fractured horizontal well. Immediately after the end of wellbore storage period, a linear flow of half slope appears*

**Table 5–2** Summary of Pre Fracturing MIT Test

| Period | Duration (hrs) | Choke size (inch) | WHP (psi) | BHP (psi) | Flow rate (MMscfd) |
|---|---|---|---|---|---|
| Initial shut-in | 136 | Closed | 2766 | 3383 | 0 |
| First flow | 3.95 | 0.375 | 2019 | 2579 | 6.69 |
| First Shut-in | 4.00 | Closed | 2749 | 3351 | 0 |
| Second flow | 4.00 | 0.5 | 1513 | 2075 | 9.25 |
| Second shut-in | 4.00 | Closed | 2736 | 3325 | 0 |
| Third flow | 4.00 | 0.625 | 1145 | 1761 | 11.14 |
| Third Shut-in | 4.00 | Closed | 2725 | 3303 | 0 |
| Forth flow | 4.00 | 0.75 | 871 | 1572 | 12.31 |
| Extended flow | 32.00 | 0.75 | 868 | 1545 | 12.04 |
| Final shut-in | 201 | Closed | 2758 | 3379 | 0 |

**Table 5–3** Summary of Post Fracturing MIT Test

| Period | Duration (hrs) | Choke size (inch) | WHP (psi) | BHP (psi) | Flow rate (MMscfd) |
|---|---|---|---|---|---|
| Shut-in | | 0 | 2757 | 3365 | 0 |
| First flow | 12.00 | 0.5 | 2092 | 2763 | 12.75 |
| First Shut-in | 12.00 | 0 | 2737 | 3314 | 0 |
| Second flow | 12.00 | 0.65 | 1739 | 2534 | 16.87 |
| Second shut-in | 12.00 | 0 | 2718 | 3277 | 0 |
| Third flow | 12.00 | 0.75 | 1419 | 2374 | 20.14 |
| Third Shut-in | 25.33 | 0 | 2715 | 3278 | 0 |
| Forth flow | 11.67 | 1 | 932 | 2200 | 24 |
| Extended flow | 48.00 | 1 | 878 | 2083 | 22.7 |
| Final shut-in | 90.00 | 0 | 2707 | 3117 | 0 |

*for a short duration of time (this line could be masked by the wellbore storage period). If the fracture conductivity is low, the line with a half slope would not exist. This flow represents the flow from the artificial Fracture. After the end of the previous stage, the adjacent reservoir begins to contribute forming quarter slopes called bilinear flow. Finally, the flow regime will represent the reservoir and the boundary condition. Table 5–4 shows the results of a pressure transient test before and after fracturing.*

The pressure transient test indicates a heterogeneous reservoir. The Kh-value (the product of permeability times the thickness of the reservoir) increased from 150 md-ft to 243 md-ft after fracturing. A negative skin value (–4) confirms that the wellbore is stimulated by the fracture treatment as indicated by the increase in deliverability (or rate production).

**Table 5–4** Results of Pressure Transient Test Analysis

| | Pre-frac | | Post-frac | |
|---|---|---|---|---|
| Well parameters | Semi log | Derivative | Semi log | Derivative |
| Final rate, MMscfd | | | 22.7 | 22.7 |
| Net pay (h), ft | 50 | 50 | 50 | 50 |
| Porosity | | | 8% | 8% |
| Wellbore radius ft | 0.26 | 0.26 | 0.26 | 0.26 |
| Formation temperature, F | 286 | 286 | 286 | 286 |
| K, md | | | 4.862 | 4.013 |
| Kh, md-ft | 150 | | 243.1 | 243.1 |
| Skin, S | | | –3.917 | –4.013 |
| Extrapolation pressure, P*, psi | 3376.33 | | 3383.42 | 3399.9 |
| C, bbl/psi | | 2.92E-02 | | 1.41E-01 |
| $C_D$ | | 382.4 | | 1845.5 |

Production data analysis provides assessment of the effectiveness of fracturing. Both the flow rates and wellhead pressures were measured on a daily basis. The wellhead pressures were converted to bottom-hole pressure by a multi-step calculation that accounts for the variation of gas density with pressure and temperature. The productivity index, which indicates the performance of the well, is the ratio of the production rate to the pressure drawdown. Table 5–5 shows the production data along with the productivity index before and after the fracture treatment for various choke sizes. After fracturing, the productivity index increased to more than 135%.

### 5.8.2 Hydraulic Fracture Design and the Use of Qualitative Data (the Eagle Ford Gas-Shale)

In a study conducted by Stegent et al. (2011), it was determined that fracturing does not stimulate all shale formations in the same manner. The completion engineer faces the challenge of designing a fracturing project with little information or tools relevant to a specific shale formation. This study provides a qualitative method that a completion engineer can use as a guideline when designing a fracturing project at a site that has little reliable historical data. In designing a

**Table 5–5** Production Data

Pre treatment flow data

| Choke size inch | WH P psi | BHP psi | Q MMscfd | PI scfd/psi |
|---|---|---|---|---|
| 0.5 inches | 1513 | 2075 | 9 | 7435 |
| 0.625 inches | 1145 | 1761 | 11 | 6910 |
| 0.75 inches | 868 | 1545 | 12 | 6373 |

Post treatment flow data

| Choke size inch | WH P psi | BHP psi | Q MMscfd | PI scfd/psi | Increase PI |
|---|---|---|---|---|---|
| 0.5 inches | 2092 | 2764 | 13 | 19173 | 158% |
| 0.625 inches | 1739 | 2534 | 17 | 16572 | 140% |
| 0.75 inches | 1419 | 2375 | 20 | 15052 | 136% |

fracturing project, consideration should be given to: (1) the type of hydrocarbon that is anticipated to be produced, (2) fracture complexity of the reservoir, (3) lithology and mineralogy of the rock, (4) geo-mechanical properties of the rock, (5) other reservoir parameters, and (6) production history (if available). The results of this study may be applied to any horizontal completion in low/ultra-low permeability reservoirs. Two main qualitative components of fracture design are reservoir characterization and design considerations.

Reservoir characterization is critical in hydraulic fracturing. Shale formations vary significantly in their properties and, therefore, understanding reservoir characteristics is vital for fracture design. A case in point is the Eagle Ford gas-shale formation, which is considered to be a hydrocarbon-generating source rock that extends laterally from southwest to northeast Texas (Figure 5–2). The Eagle Ford shale was deposited during the Cretaceous period (145 million years ago) and can be characterized as a mixture of mudstone and chalk that has been enriched with organic material. The characteristics of the Eagle Ford formation change substantially across the SW-NE strike of the play as well as from the NW-SE dip toward the Gulf of Mexico (Figure 5–4). The thickness can range from about 45 to 500 ft. Formation depths range from about 2,500 to 4,500 ft on the down-dip side from San Antonio toward the Texas coast.

Core analyses from strategic locations throughout the shale are essential for design of fracture treatments because each location presents different properties that make each fracture design unique. Thin sections of Eagle Ford cores in the northeast end of the trend characterize the formation as a planar, laminated shale with numerous bedding-plane fractures and a matrix rich in organic matter. Well logs are also taken in the vertical wells drilled into the shale formations and calibrated with the laboratory analyses of the cores. Thus reliable well-logs of the horizontal wells can be obtained later to guide the locations and number of staged fractures that are required for maximum production of the shale-gas.

Core analyses of the Eagle Ford shale uncovered the presence of about 25% clay that swells from 5–10% when in contact with fresh water throughout most of the shale bed. This means that if fresh water is used for frac-fluid, the hydrated clays would close and effectively plug both matrix pores and fracture channels, greatly reducing ultimate hydrocarbon recovery. Thus the frac-fluids had to include a 1.0 molar solution of sodium or potassium chloride (6% NaCl; 7% KCl) for protection against clay swelling.

The Eagle Ford shale is a soft rock with a Brinell hardness number of 22 and therefore the proppant can be embedded into the matrix by

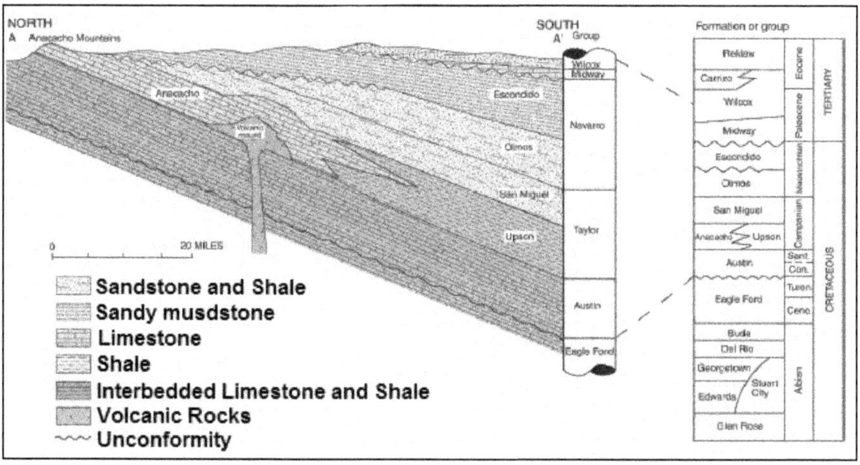


**Figure 5–4** *Cross-section view of the Eagle Ford Shale. (US Geological Survey 2010)*

**Table 5–6** Reservoir Properties of Eagle Ford Shale from Core Data

| Parameter | Unit | Range |
|---|---|---|
| Total Organic Carbon | % | 2–9 |
| Porosity | % | 8–18 |
| Water Saturation | % | 7–31 |
| Permeability | nanodarcies | 20–1,200 |
| Static Young's Modulus | pounds/square inch | 1.00E+06–2.50E06 |
| Poisson's Ratio | | 0.25–0.27 |

the closure pressure of the formation when the fracture treatment is concluded, resulting in closure of the fractures (Tiab and Donaldson, 2012). Larger mesh size proppants (20/40) were used to maintain conductivity of the fracture channels. In contrast, the Barnett gas-shale is a very hard, brittle formation (Brinell hardness number of 80) and exhibits little or no proppant embedment.

The Eagle Ford shale is rich in both oil and gas. Core geochemistry can identify the type of kerogen and its thermal maturity and determine whether the well is in a dry gas or liquid gas zone. If liquids are expected, then high conductivity proppants with larger mesh size are used. For areas that can have both oil and gas, fracture design must address simultaneous three phase flow (water, oil, and gas) and conditions that promote emulsion formation. A non-emulsifying surfactant is included in the fracture fluids. Typically, the addition of a non-emulsifying surfactant provides protection from incompatibility with the stimulation fluids and the reservoir fluid.

Completion fluid (fluid used in the final stage of fracturing) and proppants are critical to success of any fracture and therefore receive considerable attention for proper formulation. The list below details the more important petrophysical and chemical design criteria for completion fluids:

- At the beginning of the fracture treatment, both the fluid injection rate and specific design fluid viscosity of the frac-fluid are critical to create the required system of fractures and establish accurate fracture width. Real-time micro-seismic mapping is used to control the proper injection rate and physical fluid properties required to obtain effective fracture geometry.
- Ductile (soft) shale, such of that of most of the Eagle Fork requires a more viscous fracture fluid.
- Relatively shallow reservoirs' depth allow placement of high concentrations (4 lbs/gal) of large-mesh proppant (20/40 mesh) with slick water; but at deeper depths more viscous fluids are required.
- Large proppant size with higher conductivity is used for liquid gas production.
- Proppant embedment, formation fines, crushed proppant, and proppant diagenesis can all have a major impact on sustaining conductivity with time.
- Surface modifying agents can help minimize the effect of slow proppant-pack conductivity reduction with respect to time.
- Execution of the stimulation treatments requires managing vast amount of data (mud logs, bit records, cuttings analysis, core analysis, pilot wellbore open-hole log analyses, cased-hole horizontal log analyses, 3-D surface seismic surveys, and micro-seismic fracture mapping, among others). Whenever it is possible, a single software system capable of 3-D visualization of real-time analysis, geological properties, petrophysical properties, and statistical analyses is used.
- Fracture mapping using micro-seismic tools can provide critical information regarding the success (or failure) of a completion.

- Multi-well and multi-petrophysical analyses coupled to production provide valuable information and improved understanding for future fracture design.

### 5.8.3 Fracture Stimulation in the Bakken Formation (Lessons Learned)

Hlidek and Rieb (2011) conducted a study to evaluate 460 wells that were fractured in the Bakken oil shale and developed a list of best practices. The Bakken shale is situated in the northern region of the Williston Basin of North Dakota and parts of Montana and Canada at a depth of about 4,900 ft, Figure 5–5. The permeability of the Bakken formation is very low, ranging between 0.01 md to 0.5 md. The study area has a gross pay thickness varying between 16 and 96 ft. The formation temperature is about 65°C and is overlain by the Lodgepole aquifer. The matrix porosity is between 5 and 12%, and the average water saturation is 50%. Fracture treatments that grow into the aquifer are not desirable because it results in excess water production. All wells in this study are drilled horizontally and normally completed as open-holes in the Middle Bakken. Most completions consist of isolated multistage treatments. The study results provide a list of best design practices for fracturing the formation:

- Vertical wells were successfully treated using non-crosslinked linear gelled fluids at low rates to limit fracture height so that it would not extend into the aquifer. Fracturing horizontal wells require cross-linked fluids to carry the proppants. Both zirconium and borate cross-linked gels worked equally well.

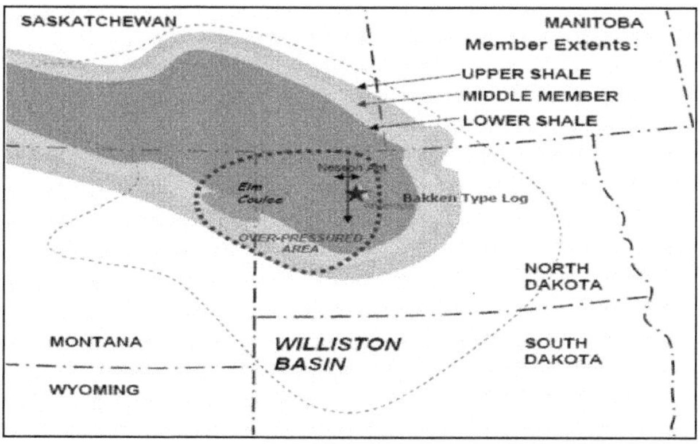


**Figure 5–5** *The Bakken Formation in northern United States and Canada. (Source: USGS)*

- Produced water can be used to prepare the injection fluid but requires modification of the chemical formulations to avoid incompatibility issues. Forty percent of all treatments in the study group utilized produced water as a base fluid. Use of produced water can lower the demand on fresh water and is of great advantage where fresh water is scarce.
- The average water production of the completed wells increased with proppant concentration.
- The fracture treatment fluid injection rate (which can define the fracture height) did not show a correlation with fluid production.
- A greater number of fractures along the length of the horizontal wells may not increase overall production.
- Well-bore orientation with respect to maximum horizontal stress did not have a significant impact on overall production.
- The advantage of open-hole completions is that production from the fracture treatment is supplemented by flow of formation fluids into open-hole (un-stimulated flow).

CHAPTER 6

# Environmental Impacts of Hydraulic Fracturing

## 6.1 Surface and Subsurface Environmental Effects

Hydraulic fracturing can have a wide range of impacts on human health and the environment. These impacts are caused by environmental aspects including activities and products related to the fracturing process. The chemicals used, and waste generated, by hydraulic fracturing operations are referred to as "product." Characteristics of the products (toxicity, volumes, duration of exposure, and quantity of the chemicals) used can have serious detrimental effects on the surface and subsurface environments.

Ecological impacts are caused by the environmental aspects associated with the hydraulic fracturing process which affect the physical environment as well as the flora and fauna of the area. Emphasis is placed on the detrimental effects of the chemicals on the physical environment.

## 6.2 Water Withdrawals

Large volumes of fresh water are required for fracturing operations. Fresh water is used to enable mixing of the chemicals with predictable reactions. Otherwise, the water must be treated extensively prior to use to ensure that precipitation of some of the chemicals does not occur. The principal concern is that fresh surface and ground waters are the source of drinking and farming water in most areas.

The percentage of water usage for hydraulic fracturing with respect to the overall water usage from a typical basin is small (DOE, 2009). However, the large amount of water required for hydraulic fracturing in a very short time can have a significant, adverse impact on the local water supply. For aquifers with low permeability, the radius from the wellbore that is affected (radius of influence) by the water withdrawal can be small but the ecological impact on this small radius can be severe. Impacts that can produce long-term alterations of the aquifer can include the introduction of oxygen in water and recharging the subsurface aquifer, which in turn can generate geochemical reactions. Recharged water trickling into the subsurface aquifer can introduce contaminants and excessive withdrawal of water can cause compaction of the aquifer, resulting in subsidence and permanent loss of water-bearing capacity (loss of matrix porosity). If water is withdrawn from streams during periods of draught, it can affect fish and other aquatic life, recreational activities, and the supply of water to municipalities and industries.

Each horizontal well that is fractured can require between two to four million gallons of fresh water (it was estimated that 35,000 wells were fractured in 2006 alone across the United States). The number is increasing with greater interest in production from gas-shale and tight sand formations. To fracture 35,000 horizontal wells requires between 70 and 140 billion gallons of fresh water, which is equivalent to the water demand of 40 to 80 cities with populations of 50,000. Almost all of the large volume of water used for hydraulic fracturing is lost forever after it is injected unless the flow-back is treated and re-introduced back into the hydrologic cycle.

In the field, to meet these large water volume requirements for fracturing, water is commonly stored in large portable steel tanks located at the well site. For multiple wells, water also can be stored in impoundment ponds. Water used to fracture the Barnett and Fayetteville shales, for example, was stored in large lined impoundments ranging in capacity from 8 to 163 million gallons to service up to 2,000 gas wells (Satterfield et al., 2008). If the impoundment ponds must be placed in locations that are remote from the hydraulic fracturing sites, the water is transferred from the impoundments to the well site by trucks or pipelines, depending upon the site-specific conditions.

The ecological impact of withdrawing and transporting large volumes of water can vary depending upon regional conditions. In arid regions such as the Bakken shale of North Dakota, where aquifer and surface waters are limited, the withdrawal of large volumes of water can cause severe ecological impacts. This can lead to lowering of the water table or dewatering of drinking water aquifers, decreased stream flows, and reduced volumes of water in surface water reservoirs. To avoid lowering the water tables of shallow aquifers, deep wells were constructed to extract water for use in the Haynesville shale area of Louisiana.

Lowering the water table can affect aquifer water quality in several ways. In coastal regions, it can cause salt water intrusions into the groundwater aquifer. When pressure is decreased in the aquifer by a large withdrawal of water, seawater can seep into the aquifer and contaminate it with high salinity. In addition, the aquifer water quality also may be affected by exposing naturally occurring minerals to an oxygen-rich environment, causing chemical changes that affect mineral solubility and mobility. For example, the lowering of water tables in Bangladesh introduced oxygen-rich water from the surface, recharging the oxidized arsenic in sediments from a less soluble arsenite to an arsenate state, which is a more water soluble compound. This resulted in the contamination of more than 4 million drinking-water wells with high levels of arsenic, producing one of the worst public health disasters in Bangladesh. Aerobic bacteria also may be introduced by infiltrating surface waters causing taste and odor problems that require treatment of the water before it can be used for human consumption. Lowering the water table also may bring about collapse of the pore space in the aquifer, as mentioned above. The compaction of pore space can be extensive enough to produce noticeable surface subsidence and damage to structures and other property on the ground surface.

Withdrawals of large quantities of water from surface water resources (streams, lakes, and ponds) can significantly affect the movement and distribution of this vital resource. It is important to recognize that ground and surface water are hydraulically connected. Therefore changes of the quality and quantity of the surface water can produce adverse effects on shallow subsurface aquifers and vice versa (Winter et al., 1998).

### 6.2.1 Mitigation Measures

Reduction of fresh water withdrawal can be achieved by treating and recycling flow-back and produced water from the fracturing operation. It is estimated that 20–75% of the frac-fluid injected may be produced back to the surface within the first two weeks after fracture completion. If all of this water is recycled, it could have a significant positive impact in reducing the demand for fresh water. On the high side, if 4 million gallons are injected, as much as 3 million gallons of flow-back water would require treatment, if all of it is to be recycled. On the low side, if 1 million gallons are injected, it could mean treatment of only 200,000 gallons of flow-back water.

Treated flow-back water is mixed with fresh makeup water before chemicals required to formulate frac-fluid are added. There are some challenges associated with reusing flow-back water, such as constructing a treatment system on-site to quickly treat a large volume of water on a temporary basis and then moving the equipment and setting it up at a new site in few weeks. The treatment equipment must include a large tank, or pond, designed to hold the maximum amount of flow-back water anticipated. Transporting such equipment to remote places in harsh terrain and treating high concentrations of total dissolved solids, as well as dissolved chemical constituents found in flow-back water, would present a costly endeavor. However, recycling flow-back water (if it is not properly treated) may require considerable dilution with fresh water.

Another alternative of reducing the demand for fresh water on local resources is to use non-potable groundwater, but non-potable water may require treatment for removal of salts that could cause precipitation when chemicals necessary for the fracture process are added.

Yet another alternative that can be effective in reducing water demand on groundwater or surface water during dry seasons is to withdraw and store surface water in impoundments during wet seasons or when water is released from a reservoir. Utilizing seasonal flow allows planning of withdrawals to avoid potential impacts to municipal drinking water supplies or to aquatic or riparian communities. For the Fayetteville shale of Arkansas, an impoundment capable of storing 163 million gallons of water provided water for the fracturing operation. Water was withdrawn from the Little Red River during periods of high flow (storm events or hydroelectric power generation releases from Greers Ferry Dam upstream) when excess water is available. The project included careful monitoring of the stream water quality.

## 6.3 Surface Spills

Hydraulic fracturing operations are intensive over a short period of time, usually a few weeks , and require a large amount of equipment and chemicals, such as: pumps, proppants, vehicles, and other equipment that can result in unintentional spills on the surface. These spills can seep into shallow groundwater aquifers, flow into surface waters, evaporate into the air, or stay on the surface soil depending upon the type of spill and its location. Surface spills can occur as a result of a host of unpredictable accidents, such as: tank ruptures, equipment or surface impoundment failures, overfills, vandalism, accidents, ground fires, or improper operations.

Chemical compositions of spills can be extremely varied because of the large inventory of different chemicals that are used for fracturing. It largely depends upon the type of fracturing fluid designed for a particular location. The US EPA has identified hundreds of chemicals that are used as additives in frac-fluids (US EPA, 2011). It is estimated that the concentration of chemicals in the frac-fluids is between 0.5% and 2%. This may seem like a small quantity, but in reality their volumes are significant because of the large volume of frac-fluid that is generally used for each fracture treatment of a horizontal well (1 million to 4 million gallons). This means that the volume of chemicals used can vary between 5,000 and 80,000 gallons for every horizontal well. Some of the chemicals are hazardous to human health, even in very small quantities, while others are benign. This large volume of chemicals has to be managed properly, making sure that it does not spill and contaminate the environment.

The pathway by which spilled chemicals may migrate to groundwater, surface water, and air depends on many factors, including the site, type of chemicals, and/or fluid properties. Site-specific factors may include the location of the spill with respect to the ground and surface water resources, weather conditions at the time of the spill, and the type of surface on which the spill occurred. Chemical-specific factors include the chemical and physical properties of the chemicals such as vapor pressure, density, solubility, diffusion, and partition coefficients. These properties govern how easily the chemicals can migrate from one medium (surface soil) to another (groundwater, surface water, or air).

### 6.3.1 Mitigation Measures

Surface spills and their effects can be minimized by: (1) providing adequate training for the crew handling equipment and chemicals; (2) using chemicals that are not toxic to the environment and are biodegradable; (3) using appropriate liners to contain spills; (4) using double-walled tanks to minimize accidents related to rupture of single walls; (5) having site-specific spill prevention control, countermeasure plans, and the associated equipment/chemicals necessary to neutralize any spill; and (6) proper housekeeping.

## 6.4 Wastewater Management

Wastewater management and disposal laws forbid operators from directly discharging wastewater associated with shale-gas production into waterways. The two options primarily used today to manage wastewater are underground disposal wells and recycling. Lesser used options include wastewater treatment prior to discharge into public waterways and evaporation in open storage ponds.

Wastewater generated from fracturing operations is commonly injected into deep formations using injection wells. Disposal of wastewater into these formations is possibly the lowest cost option for disposal and the process has been carried out safely for a long time. However, not all regions have appropriate formations where disposal is possible. In Texas, the Barnett shale wastewater can be injected into permeable rocks more than a mile underground, but the Marcellus shale region apparently does not have a suitable disposal formation for waste injection. In addition, state and federal regulations present strict guidelines for construction and operation of deep injection wells but do not restrict injection near fault zones, where injection of large volumes of wastewater can lubricate zones under stress and result in minor earth tremors (less than 4 on the Richter scale). So far, the regulations for disposal wells have focused on protecting aquifers, not preventing seismic activity. In December 2010, the Arkansas Oil and Gas Commission imposed a moratorium on new wastewater disposal wells in an area that had begun experiencing minor earthquakes. In March 2011, the Commission asked operators to shut down wastewater disposal wells close to a fault when an earthquake of magnitude 4.7 occurred. The Commission also placed a moratorium on new disposal wells over a 1,100 square mile area of the zone. In Ohio, where companies dispose of shale-gas wastewater from Ohio and Pennsylvania, government officials stopped operation of a disposal well in January 2012 and delayed construction of four wells after eleven earthquakes, including one of magnitude 4.0 occurred near Youngstown.

### 6.4.1 Mitigation Measures

General preventive measures to ensure against contamination from fracture treatment wastewater include the use of secondary containments (tanks or ponds), mats, catchments, and groundwater monitors, as well as the establishment of buffer zones around surface waters.

When designing an underground disposal well, it is essential that the geology of the area be evaluated closely by a seismic survey to ensure that the injection well is not close to a geological fault zone that can slip and cause earthquakes. The subsurface formation selected for injection of wastewater should have high porosity and permeability, and there should be impermeable zones above the injection zone to offer protection to any shallow water aquifers. In sensitive areas, monitoring wells can be installed near the injection well to detect any leaks or migration of the injected fluid into the aquifers.

## 6.5 Air Emissions

Temporary emission of hydrocarbons from shale-gas wells can occur when fractured wells are being prepared for production. During the flow-back period, which lasts only a few days, spent frac-fluids, formation brine, and gases (methane, ethane, and other volatile hydrocarbons) are produced from the formation at high velocity and considerable volume. The effluent from the well is piped to a gas/liquid separator to remove the liquids and store the gas for compression and transport. In some cases, the gas may be burned from a flare to eliminate the excess gas.

Other processes and equipment during fracturing also can emit transient air pollutants. These include: pumps, compressors, generators, and diesel-fueled trucks, as does any construction or heavy industrial activity. Particulate matter may be released from dust and soil, or soot from vehicle diesel exhaust. Ozone itself is not released directly during fracturing operations, but two of its main precursors, volatile organic compounds and oxides of nitrogen, may combine with sunlight to form ground-level ozone.

Air-quality impacts from oil and gas operations in Texas, Wyoming, Colorado, and Utah are well documented; however, any contribution from hydraulic fracturing should be relatively small.

### 6.5.1 Mitigation Measures

Companies are using reduced emission completions (also known as "reduced flaring completions" or "green completions"). Portable equipment is available for on-site separation of gas from the liquids and solids produced during the period of high-rate flow-back, followed by compression and transported in pipelines, as mentioned above.

Additionally, engine manufacturers are constantly improving their technology to reduce the amount of nitrogen oxides emission from their engines. One key method has been the use of catalytic converters to chemically transform the nitrogen oxides into inert compounds. The addition of catalytic emission controls has successfully lowered engine emissions from 20 grams of nitrogen oxides per horsepower-hour to 2 grams of nitrogen oxides per horsepower-hour, or less. Also, the addition of air-to-fuel ratio-controllers reduce detrimental emission performance from these engines.

Best management practices have also been developed by the petroleum industry to reduce, or completely eliminate, adverse air emissions in the field. Some of them are:

- Replacement of high-bleed pneumatic devices (transducers, valves, controllers, etc.) with low-bleed devices to reduce emission of gas.
- Use of infrared cameras in the field to visually identify any hydrocarbon leaks so that they can be repaired rapidly to reduce the ecological impact. The cameras are tuned to the wavelengths that are reflected by hydrocarbon gases, so that the normally invisible gases actually become visible as "smoke" in the camera image, thus allowing companies to quickly detect and repair leaks.
- Installation of flash-tank separators for situations that require the use of dehydrators. They can recover 90–99% of the methane that would otherwise be flared or vented into the atmosphere.

## 6.6 Water Impoundments

Water impoundments are used to store water for preparing frac-fluids. They also can be used to store flow-back water before and after treatment. In arid regions, the industry uses open pits and tanks to evaporate liquid from the solid pollutants. Full evaporation ultimately leaves precipitated solids that must be disposed in a landfill. These solids are regulated under the Resource Conservation and Recovery Act, Subtitle D, and classified as non-hazardous waste. States usually require pits to be built to specifications that include ground compaction, multiple, heavy wall liners, monitoring methods to detect leakage, and storm water control.

In fall 2011, some wastewater ponds in Pennsylvania overflowed because of excessive rainfall from tropical storm Lee. There also is some concern that evaporative pits may allow air emissions of volatile organic compounds and other pollutants. Birds and wildlife, and sometimes domesticated animals like cattle, can mistake these pits for freshwater sources.

### 6.6.1 Mitigation and Innovation

The industry is increasingly replacing open pits with closed-loop fluid systems that keep fluids within a series of pipes and watertight tanks inside secondary containment. Some states, such as New York, are proposing to ban open pits. Additional measures include establishing setback requirements for open pits, determining the composition of wastewater stored in evaporative ponds for appropriate disposal or treatment, since contaminants can become more concentrated as water evaporates, and placing a fence around open pits to keep out animals.

## 6.7 Human Health Impacts

The materials and waste associated with hydraulic fracturing have the potential to impact human health. The potential impacts depend primarily on the types of material released, its concentration, and the duration of exposure. The effect on the biotic community to which it is released could be minimal in the case of low concentrations, or low toxicity, of the compounds, or it might be significant.

Toxicity occurs when a material causes a deleterious effect on an organism, population, or community following exposure to a substance. Toxicity is the degree to which a chemical or a substance can damage an organism and is indicated by how it impacts the life and health of living organisms. Chemical characteristics, such as structure, water solubility, vapor pressure, partition coefficients can be used to assess the toxicity of chemicals.

Measurement of toxicity is commonly conducted by assessing the results of the size of a dose and concentration of the chemical compound. The *dose* is the amount of a substance that has been absorbed into the tissue of a test species, while the *concentration* is the quantitative measure of a substance in the environment in which the species lives (for example: milli-grams/cubic meter of air, or milligrams/cubic centimeter of liquid, etc.). In addition, the interval of time of exposure is essential information for measurement of toxicity (Reis J.C., 1996).

When a chemical substance is released in an open area, such as a hydraulic fracturing site, it becomes a part of the local environment. Such a release does not always lead to human or animal exposure since direct contact with the contaminant is required. Exposure to the chemical may occur by inhalation, eating (oral), drinking, or by skin (dermal) contact. There has apparently been considerable public concern regarding the toxicity of chemicals used in hydraulic fracturing fluids (US EPA, 2011). In response to these concerns, the US House of Representatives Committee on Energy and Commerce conducted an investigation to examine the practice of hydraulic fracturing in the United States. The investigation revealed that between 2005 and 2009, the 14 leading oil and gas service companies used more than 2,500 hydraulic fracturing products containing 750 chemicals and other components. This included twenty-nine chemicals that are: (1) known, or possible, human carcinogens; (2) regulated under the Safe Drinking Water Act for their risks to human health; and (3) listed as hazardous air pollutants under the Clean Air Act (Waxman et al., 2011).

Toxicological effects are caused by the chemicals used in the hydraulic fracturing processes and constituents of the flow-back. Although some substances are toxic in high concentrations, they may be essential for normal biological processes in low concentrations, such as trace minerals and minute quantities of heavy metals. For most chemicals, there exists a threshold dose below which the chemical does not have a toxic effect. The threshold dose depends on the ability of the body to detoxify and excrete the substance, as well as repair any damage through normal biological processes; if, however, the body is exposed to a dose that is higher than can be repaired by normal biological processes, then toxic impact will occur.

For carcinogenic and mutagenic substances, however, it is assumed that there is no threshold dose; that is, the chemical is toxic even in very small doses (US EPA, 2005). A mutagenic substance can change the rate of mutation, while a carcinogenic substance can cause cancer in the body. Many substances, called mutagens, can alter the structure of the DNA molecules in individual cells. Most mutations result in the death of the individual cells effected, with no reproduction of the mutation. If a mutated cell survives and results in future birth defects, the substance is called teratogenic. If the mutation results in cancer, the substance is called carcinogenic. As a rule, nearly all carcinogens are also mutagens, but not all mutagens are carcinogens.

When a chemical is released into the environment, it does not always lead to exposure, and without exposure the chemical will not have an impact on the human health. Even when a person is exposed to a toxic chemical, many factors determine whether that person will be harmed. These factors include dose (quantity) and duration. Other factors like age, sex, genetic traits, lifestyle, and state of health also play role in defining the risk of adverse impact to human health.

One of the concerns about adverse health effect is that some chemicals generated by hydraulic fracturing are toxic, even at a very low concentration, because they disrupt normal hormonal functions. According to the "low dose hypothesis," health effects occur at doses far below levels previously determined to be safe using well-established toxicological procedure and principles.

## 6.8 Where to Get Toxicity Information

The EPA's Integrated Risk Information System is a human health assessment program that evaluates scientific information on effects that may result from exposure to environmental contaminants (US EPA, 2002). The EPA's risk information system provides the highest quality science-based human health assessments to support the Agency's regulatory activities (US EPA, 2005). The program provides the health profiles of chemicals to which the public may be exposed from releases to air, water, and land caused by the use and disposal of chemicals.

The integrated risk information system summaries also may include references to oral and inhalation doses that result from chronic, and other exposure durations, as well as carcinogenic assessments. The oral and inhalation dose values provide quantitative information for use in risk assessments for health effects that are known, or assumed to be produced, from exposure (US EPA, 2005; US EPA 2011).

The oral dose, expressed in units [(mg/kg)/day; weight of chemical/body weight per day], is defined as an estimate (with uncertainty spanning perhaps an order of magnitude) of daily exposure to a general human population that is likely to be without an appreciable risk of deleterious effects during a lifetime (US EPA, 2006). The inhalation dose, expressed in units of $mg/m^3$, is similar to the oral dose, which provides a continuous inhalation exposure estimate. The inhalation dose includes toxic effects for both the respiratory system and for effects that are peripheral to the respiratory system (systemic effects). Thus the biological effects of inhaled vapors depend on vapor uptake of the lungs and the subsequent distribution of the toxic substance to the rest of the body (Asgharian et. al, 2012).

Reference values are derived for chronic exposure durations as follows: (1) acute, less than 24 hours (2) short term/subacute, less than 30 days, and (3) subchronic, one to three months.

An example excerpt from the EPA Integrated Risk Information System website is presented on page 128, where BMDL represents the Bench Mark Dose Level; UF stands for Uncertainty Factor; and MF stands for Modifying Factor (US EPA, 2002). The chosen chemical is benzene, which is sometimes used as a constituent of hydraulic fracturing fluid.

The carcinogenicity assessment provides information on the carcinogenic hazard potential of the substance in question, and quantitative estimates of risk from oral and inhalation exposure. Quantitative risk estimates may be derived from the application of a low dose extrapolation procedure (Ksoy, 1989; Keller and Snyder, 1986).

**I. Chronic Health Hazard Assessments for Non-carcinogenic Effects**

**I.A. Reference Dose for Chronic Oral Exposure (RfD)**

Substance Name:Benzene
CASRN:71-43-2
Last Revised:04/17/2003

The oral Reference Dose (RfD) is based on the assumption that thresholds exist for certain toxic effects such as cellular necrosis. It is expressed in units of (mg/kg)/day. The US EPA has evaluated this substance for potential human carcinogenicity. UF is Uncertainty Factor; MF is Modifying Factor.

**I.A.1.** Oral RfD Summary

| Critical Effect | Experimental Doses* | UF | MF | RfD |
|---|---|---|---|---|
| **Decreased lymphocyte count (Human occupational inhalation study; Rothman et al., 1996)** | **BMDL = 1.2 mg/kg/day** | **300** | **1** | **4.0 ? $10^{-3}$ mg/kg/day** |

## 6.9 Chemicals Present In Hydraulic Fracturing Fluid

Based on congressional data gathered from oil companies, there were 750 chemicals used for fracturing operations (US EPA, 2011). As part of New York State's Draft Supplemental Generic Environmental Impact Statement related to "Horizontal Drilling and High-Volume Hydraulic Fracturing in the Marcellus Shale," the Department of Environmental Conservation compiled a list of chemicals and additives used during hydraulic fracturing (William, 2012). Table 6–1 shows various types of chemicals proposed for hydraulic fracturing in New York. Chemicals in brackets have not been proposed for use in the state but are known to be used in other states or shale fracture treatments. Table 6–2 is a compilation of the chemicals used in hydraulic fracturing and their hazard characteristics.

**Table 6–1** Chemicals Proposed for Hydraulic Fracturing in the State of New York

| Additive Type | Description of Purpose | Example of Chemical |
|---|---|---|
| Proppant | "Props" open fractures and allows gas / fluids to flow more freely to the well bore. | Sand [Sintered bauxite; zirconium oxide; ceramic beads] |
| Acid | Cleans up perforation intervals of cement and drilling mud prior to fracturing fluid injection, and provides accessible path to formation. | Hydrochloric acid (HCl, 3% to 28%) or muriatic acid |
| Breaker | Reduces the viscosity of the fluid in order to release proppant into fractures and enhance the recovery of the fracturing fluid. | Peroxydisulfates |
| Bactericide/ Biocide | Inhibits growth of organisms that could produce gases (particularly hydrogen sulfide) that could contaminate methane gas. Also prevents the growth of bacteria which can reduce the ability of the fluid to carry proppant into the fractures. | Gluteraldehyde; 2-Bromo-2-nitro-1,2-propanediol |
| Buffer/pH Adjusting Agent | Adjusts and controls the pH of the fluid in order to maximize the effectiveness of other additives such as crosslinkers. | Sodium or potassium carbonate; acetic acid |
| Clay Stabilizer/ Control | Prevents swelling and migration of formation clays which could block pore spaces thereby reducing permeability. | Salts (e.g., tetramethyl ammonium chloride) [Potassium chloride] |
| Corrosion Inhibitor | Reduces rust formation on steel tubing, well casings, tools, and tanks (used only in fracturing fluids that contain acid). | Methanol; ammonium bisulfate for Oxygen Scavengers |
| Crosslinker | The fluid viscosity is increased using phosphate esters combined with metals. The metals are referred to as crosslinking agents. The increased fracturing fluid viscosity allows the fluid to carry more proppant into the fractures. | Potassium hydroxide; borate salts |

**Table 6–1** Chemicals Proposed for Hydraulic Fracturing in the State of New York (cont'd)

| Additive Type | Description of Purpose | Example of Chemical |
|---|---|---|
| Friction Reducer | Allows fracture fluids to be injected at optimum rates and pressures by minimizing friction. | Sodium acrylate-acrylamide copolymer; polyacrylamide (PAM); petroleum distillates |
| Gelling Agent | Increases fracturing fluid viscosity, allowing the fluid to carry more proppant into the fractures. | Guar gum; petroleum distillate |
| Iron Control | Prevents the precipitation of carbonates and sulfates (calcium carbonate, calcium sulfate, barium sulfate) which could plug off the formation. | Ammonium chloride; ethylene glycol; polyacrylate |
| Solvent | Additive which is soluble in oil, water & acid-based treatment fluids which is used to control the wettability of contact surfaces or to prevent or break emulsions. | Various aromatic hydrocarbons |
| Surfactant | Reduces fracturing fluid surface tension thereby aiding fluid recovery. | Methanol; isopropanol; ethoxylated alcohol |

**Table 6–2** Characteristics of Undiluted Chemicals Found in Hydraulic Fracturing Fluids (Based on MSDSs)

| Product | Chemical Composition Information* | Hazards Information |
|---|---|---|
| Linear gel delivery system | 1. 30–60% by wt. Guar gum derivative<br>2. 60–100% by wt. Diesel | Harmful if swallowed<br>Combustible |
| Water gelling agent | 1. 60–100% by wt. Guar gum derivative<br>2. 5–10% by wt. Water<br>3. 0.5–1.5% by wt. Fumaric acid | May be mildly irritating to eye |
| Linear gel polymer | 1. <2% by wt. Fumaric acid<br>2. <2% by wt. Adipic acid | Flammable vapors |

**Table 6–2** Characteristics of Undiluted Chemicals Found in Hydraulic Fracturing Fluids (Based on MSDSs) (cont'd)

| Product | Chemical Composition Information* | Hazards Information |
|---|---|---|
| Linear gel polymer slurry | 1. 30–60% by wt. Diesel oil #2 | Causes irritation if swallowed |
| Crosslinker | 1. 10–30% by wt. Boric acid<br>2. 10–30% by wt. Ethylene glycol<br>3. 10–30% by wt. Monoethanolamine | Harmful if swallowed<br>Combustible |
| Crosslinker | 1. 10–30% by wt. Sodium tetraborate decahydrate | May be mildly irritating to eye |
| Foaming Agent | 1. 10–30% by wt. Isopropanol<br>2. 10–30% by wt. Salt of alkyl amines<br>3. 1–5% by wt. Diethanolamine | Harmful if swallowed<br>Highly flammable |
| Foaming Agent | 1. 10–30% by wt. Ethanol<br>2. 10–30% by wt. 2-Butoxyethanol<br>3. 25–55% by wt. Ester salt<br>4. 0.1–1.0% by wt. Polyglycol ether<br>5. 10–30% by wt. Water | Harmful if swallowed or absorbed through skin |
| Add treatment—hydrochloric acid | 1. 30–60% by wt. Hydrochloric acid | May cause eye, skin, and respiratory burns |
| Add treatment—formic acid | 1. 85% by wt. Formic acid | May cause mouth, throat, stomach, skin and respiratory burns |
| Breaker fluid | 1. 40–100% by wt. Diammonium peroxidisulfate | May cause respiratory tract, eye, or skin irritation<br>Harmful if swallowed |
| Microbicide | 1. 60–100% by wt. 2-Bromo-2-nitrol, 3 propanedol | May cause eye, or skin irritation |

**Table 6–2** Characteristics of Undiluted Chemicals Found in Hydraulic Fracturing Fluids (Based on MSDSs) (cont'd)

| Product | Chemical Composition Information* | Hazards Information |
|---|---|---|
| Biocide | 1. 60–100% by wt. 2,2-Dibrom 3-nitrilopropionide<br>2. 1.5% by wt. 2-Bromo-3-nitrilo-propionamide | Causes severe burns<br>Harmful if swallowed<br>May cause allergic reaction |
| Add corrosion inhibitor | 1. 30–60% by wt. Methanol<br>2. 5–10% by wt. Propargyl alcohol | May cause eye, or skin irritation<br>May be fatal if swallowed |
| Add corrosion inhibitor | 1. 30–60% by wt. Pyridinium, 1-(phenylmethyl)-, ethyl methyl derivatives, chlorides<br>2. 15% by wt. Thiourea<br>3. 5–10% Propan-2-ol<br>4. 1–5% by wt. Poly (oxy-1,2-eth-anediyl)-noniphenyl-hydtoxy<br>5. 10–30% by wt. Water | Cancer hazard<br>Causes severe burns to respiratory tract, eyes, skin<br>Harmful if swallowed or absorbed through skin |

- Data extracted from Table 4-1, page 4-9 of Chapter 4, Hydraulic Fracturing Fluids, EPA 816-R-04-003

* Information presented is for the pure product, which is significantly diluted prior to injection. MSDS chemical composition percentages may total more than 100%

## 6.10 Chemicals Present in Flow-Back Fluids

Flow-back fluids can be composed of injected fracturing fluid, naturally occurring materials present in the formation, and any reaction, or degradation products, formed during the hydraulic fracturing process. The composition of flow-back and produced water has been shown to vary with location and time (US EPA, 2011). In general, analyses of flow-back from various reports show that concentrations of total dissolved solids can range from approximately 1,500 to more than 300,000 mg/L (Gaudlip and Paugh, 2008). The Appalachian basin tends to produce one of the higher concentrations of total dissolved solids: 250,000 mg/L (Breit, 2002). Along with high total dissolved solids values, flow-back can have high concentrations of several ions (e.g., barium, bromide, calcium, chloride, iron, magnesium, sodium, strontium, bicarbonate), with concentrations of calcium and strontium sometimes reported to be as high as thousands of milligrams per liter (Vidic, 2010). Flow-back also can contain radionuclides, which are naturally occurring radioactive materials that tend to accumulate in shale-type sediments. Some flow-back solutions from the Marcellus shale formation have been measured at concentrations as high as 18,000 picocuries per liter (pCi/L) and elsewhere in the United States above 10,000 pCi/L (Rowan., 2011; USGS, 1999). Volatile organic compounds, including but not limited to benzene, toluene, xylenes, and acetone, also have been detected. In addition, flow-back has been reported to have variations of pH ranging from 5 to 8 (Hayes, 2009).

Table 6–3 (page 140) shows chemicals typically produced during flow-back along with their health effects, recommended engineering controls, and protective equipment to handle those chemicals.

## 6.11 Regulations for Protecting the Environment

Regulations related to fracturing are changing along with the innovations and developments taking place in the petroleum industry. As the federal government has exerted limited oversight of oil and gas development, including shale-gas development, most of the authority to regulate these operations is vested at the state level. Recently, the federal government has announced several proposals to increase its oversight of shale-gas related activities. Presently, an ongoing debate exists over whether the states or the federal government should be taking the lead in overseeing shale-gas operations. Those in favor of state-led oversight argue that state authorities are better positioned to address issues concerning unique geological and hydrological characteristics and other local factors that vary significantly throughout the country. State regulators also typically have on-site experience with fracturing sites in their area. Others would like more natural gas activities to be regulated under federal law to maintain uniform standards throughout the country. Since geology, groundwater, surface water, and air do not end at state boundaries, action in one state can affect the environment in another state. Despite these debates, it seems likely that states will continue to take the regulatory lead regarding shale-gas development. In October 2011, the Groundwater Protection Council (William, 2012) counted 32 gas-producing states. Most relevant states either are reviewing or revising their regulations and permit requirements. Several nonprofit organizations are working to strengthen state regulations. The State Review of Oil and Natural Gas

Environmental Regulations assists states in documenting environmental regulations associated with natural gas exploration, development, and production. Additional working groups include the Interstate Oil and Gas Compact Commission and the Groundwater Protection Council, which in addition to overseeing the FracFocus disclosure website has developed a Risk Based Data Management System that helps states collect and publicly share data associated with their oil and gas regulatory programs.

## 6.12 Regulations Currently Applicable

The EPA administers most of the federal laws applicable to hydraulic fracturing, although development on federally owned land is managed primarily by the Bureau of Land Management, which is part of the Department of the Interior, and the US Forest Service, which is part of the Department of Agriculture. In addition, each state in which oil and gas is produced has its own regulatory agency that permits wells and oversees environmental activities and discharges, including water management and disposal, waste management and disposal, air emissions, underground injection, wildlife impacts, surface disturbance, and workers' health and safety. Many of the federal laws are implemented by the states under agreements and plans approved by the appropriate federal agencies. Some of the major regulations that affect the hydraulic fracturing process are discussed below (US DOE, 2009).

### 6.12.1 Clean Water Act

The Clean Water Act is the primary federal law in the United States governing pollution of surface water. It was established in 1970 to protect water quality and includes regulation of the pollutant limits of discharged water from oil and gas related drilling. This is conducted through the National Pollutant Discharge Elimination System permitting process; although the EPA sets national standards at the federal level, states and tribal governments can acquire primacy for the system by meeting the EPA's primacy requirements. The pollutant discharge elimination system permit program controls discharges from point sources that are discrete conveyances, such as pipes or man-made ditches. Industrial, municipal, and other facilities such as shale-gas production sites or commercial facilities that handle the disposal or treatment of shale-gas produced water must obtain permits if they intend to discharge directly into surface waters. Most oil and gas production facilities with related discharges are authorized under general permits, rather than individual permits, because there are typically numerous sites with common discharges in a geographic area.

Under the Clean Water Act, the EPA has implemented pollution control programs such as setting wastewater standards for the industry. It has also set water quality standards for a variety of contaminants in surface waters. The Clean Water Act also includes a program to control storm water discharges that requires storm water permits for municipal, industrial, and construction site storm water runoff. The permitting requirement precludes oil and gas exploration, production or treatment operations, or transmission facilities, if the runoff is not contaminated with oil, grease or hazardous substances. While the EPA storm water permitting rule contains a broad exclusion for the oil and gas industry, it is important to note that individual states and Indian tribes may still regulate storm water associated with these activities.

### 6.12.2 Safe Drinking Water Act

The Safe Drinking Water Act was passed to protect public health by regulating the nation's public drinking water supply. The law requires many actions to protect drinking water and its sources, including rivers, lakes, reservoirs, springs, and groundwater wells. The safe drinking water act authorizes the EPA to set national health-based standards for drinking water to protect against both naturally occurring and man-made contaminants that may be found in drinking water. As one aspect of the protection of drinking water supplies, the act establishes a framework for the Underground Injection Control program to prevent the injection of liquid wastes into underground sources of drinking water. The EPA and states that implement the injection control program, which sets standards for safe waste injection practices, bans certain types of injection altogether. The program provides these safeguards so that injection wells do not endanger drinking water sources.

The US EPA currently groups underground injection wells into five classes for regulatory control purposes, and has a sixth class under consideration. Each class includes wells with similar functions, constructions, and operating features so that technical requirements can be applied consistently to the class. Most injection wells associated with oil and gas activities are Class II wells. These wells may be used to inject water and other fluids, such as liquid carbon dioxide, into oil and gas bearing zones to enhance recovery, or they may be used to dispose of produced water. The EPA's Underground Injection Program is designed to prevent contamination of water supplies by setting minimum requirements for state programs. The basic premise is to prevent contamination of drinking water by keeping injected fluids within the intended injection zone. Hence the programs have established procedures for sitting, construction, operation, maintenance, monitoring, testing, and, finally, closure of injection wells.

### 6.12.3 Oil Pollution Act (Spill Prevention Control and Countermeasures)

The Oil Pollution Act is designed to reduce damage to natural resources from oil spills. The act provides for spill prevention requirements, spill reporting obligations, and spill response planning. It regulates the prevention of and response to accidental releases of oil and hazardous substances into navigable waters, adjoining shorelines, or natural resources belonging to or managed by the United States. An important component of the strategy to prevent oil spills from reaching the nation's waters is the Oil Spill Prevention, Control and Countermeasure plan. The plan is mandatory for facilities with a storage capacity greater than 1,320 gallons aboveground (or greater than 42,000 gallons below ground) and there is a reasonable expectation of a discharge of oil. According to the regulation:

- Spill Prevention Control and Countermeasure Plans must be prepared, certified by a professional engineer, and implemented by facilities that store, process, transfer, distribute, or use drills to produce or refine oil.
- Owners of facilities must establish procedures and methods and install proper equipment to prevent an oil release.
- Facilities must train personnel to properly respond to an oil spill by conducting drills and training sessions.
- Facilities must also have a plan that outlines steps to contain, clean up, and mitigate any effects that an oil spill may have on waterways.

A Spill Prevention Control and Countermeasure Plan is a site-specific document that describes and measures the controls that the facility owner has taken to prevent oil spills, and the measures that are in place, if needed, to contain and clean spills. It also includes information about the facility, the oil storage containment, inspections, a site diagram with locations of tanks (above and below ground) and drainage, and other pertinent details. Prevention measures include secondary containment around tanks and other oil-containing equipment.

### 6.12.4 Clean Air Act

The Clean Air Act is the primary means by which the EPA regulates potential emissions that could affect air quality. The EPA regulates the pollutants by developing human health-based and/or environmentally and scientifically based criteria for setting permissible levels. Geographic areas that do not meet the EPA's standards for a given pollutant are designated as "non-attainment areas." This is the case for the Barnett shale, much of which is located in or near the Dallas-Fort Worth ozone non-attainment area. As a result, Barnett shale production activities must often comply with much more stringent regulations than similar operations outside of a non-attainment area.

The EPA has established National Emission Standards for Hazardous Air Pollutants, which are national standards to control specific air emissions. In 2007, the EPA implemented a new standard referred to as the Maximum Achievable Control Technology standard for hazardous air pollutants that targeted small area sources such as shale-gas operations located in areas near larger populations. These standards limit the hazardous air pollution emissions (primarily benzene) from process vents on glycol dehydration units, storage vessels with flash emissions, and equipment leaks. Another rule requires close monitoring, record keeping, maintenance, and testing of internal combustion engines for air emissions.

Air permits are legal documents that facility owners and operators must abide by. The permit specifies the type of construction allowed, the emission limits that must be met, how the emission source(s) must be operated, and what conditions (specifying monitoring, record keeping, and reporting requirements) must be maintained to assure ongoing compliance. Shale-gas producers may need air-quality permits for a number of emission sources, including gas compressor engines, glycol dehydrators, and flares.

On April 17, 2012, the EPA issued a set of comprehensive regulatory standards for the oil and gas industry under the Clean Air Act, requiring the reduction of emissions of volatile organic compounds and methane from sources in the industry, including the hydraulic fracturing of horizontal natural gas wells drilled or hydraulically re-fractured after August 23, 2011. New wells drilled prior to January 1, 2015, must reduce volatile organic compound emissions either by using completion combustion devices or by capturing gas flow-back using reduced emission completions (also known as "green completions") with a device to control gas not suitable for entry into the flow line. A completion combustion device burns the gas that would otherwise escape during the well-completion period and usually involves flaring, unless state or local laws prohibit such combustion or require more stringent controls. A green completion uses equipment to separate gas and liquid hydrocarbons from the flow-back that occurs as the well is being prepared for production, and it may include gas gathering lines or collected system, sand traps, surge vessels, and separator tanks. Beginning January 1, 2015, owners and operators of newly fractured, or re-fractured, wells must capture the gas (not flare it) and make it available for use or sale using green completions.

### 6.12.5 Resource Conservation and Recovery Act

The Resource Conservation and Recovery Act was passed to address the growing problems of increasing volumes of municipal and industrial waste. The Act established goals for protecting human health and the environment, conserving resources, and reducing the amount of waste. Subtitle C of the Act established a federal program to manage hazardous wastes from "cradle to grave" to ensure that hazardous waste is handled in a manner that protects human health and the environment. In 1988, the EPA issued a regulatory determination stating that control of oil and gas exploration and production wastes under Resource Conservation and Recovery Act, Subtitle C, was not warranted. The EPA made this determination because it found that other state and federal programs would protect human health and the environment more effectively. In lieu of regulation under Subtitle C, the EPA implemented a three-tier strategy to ensure that the environmental and programmatic issues were addressed:

- Improve other federal programs under existing authorities;
- Work with states to improve some programs; and
- Work with Congress to develop any additional statutory authorities that may be required.

These wastes have remained exempt from Subtitle C regulations, but this does not preclude these wastes from control under state regulations or other federal regulations. The exemption applies only to the federal requirements of the Resource Conservation and Recovery Act, Subtitle C. A waste that is exempt from Subtitle C regulation might be subject to more stringent, or broader, state hazardous and non-hazardous waste regulations and other state and federal program regulations.

### 6.12.6 Fracturing Responsibility and Awareness of Chemicals Act

In response to recent public concerns about hydraulic fracturing, Congress introduced the Fracturing Responsibility and Awareness of Chemicals Act that would amend the Safe Drinking Water Act to regulate hydraulic fracturing under the same laws and regulations that are used for the Underground Injection Control program. It would also require disclosure of chemicals used in frac-fluids.

### 6.12.7 Green House Gas Reporting Rules

In a recent development, gas companies will begin reporting additional greenhouse gas emission data to the EPA by September 2012. In November 2010, the EPA finalized reporting requirements for the oil and natural gas industry under its Greenhouse Gas Reporting program. The 2010 ruling expanded the scope of existing reporting requirements to include fugitive and vented greenhouse gas emissions beginning in January 2011.

### 6.12.8 Guidance for use of Diesel Fuel as a Frac-Fluid

The EPA has authority to regulate fracturing when diesel fuel is used as the base frac-fluid and is developing permit guidance for oil and gas hydraulic fracturing activities. The draft guidance outlines requirements and technical recommendations for permitting and a description of diesel fuels for EPA underground injection control permitting. While this guidance undergoes public notice and comment, decisions about permitting hydraulic fracturing operations that use diesel fuels will be made on a case-by-case basis. After the guidance is finalized, the EPA will then have to develop implementation rules in states such as Pennsylvania and New York, where it implements the Underground Injection Control program. The 33 states that have primary enforcement authority for the Underground Injection Control program also will have to develop their own rules to implement the EPA guidance.

**Table 6–3** Typical Chemicals of Flow-Back Fluid and Health Effects

**1,6-Hexanediamine** CAS# 124-09-4

| | | | | |
|---|---|---|---|---|
| NFPA/HMIS Rating | Health: 3 | Flammability 1 | Reactivity 0 | Personal Protection J |

Toxicity—The substance is toxic to blood, kidneys, lungs, liver. Repeated or prolonged exposure to the substance can produce target organs damage. Repeated exposure of the eyes to a low level of dust can produce eye irritation. Repeated skin exposure can produce local skin destruction, or dermatitis. Repeated inhalation of dust can produce varying degree of respiratory irritation or lung damage. Repeated or prolonged inhalation of dust may lead to chronic respiratory irritation.

| | | | | |
|---|---|---|---|---|
| Possible Effects | **Carcinogenic:** NA | **Mutagenic:** NA | **Teratogenic:** NA | **Developmental:** NA |

**Engineering Controls:** Use process enclosures, local exhaust ventilation, or other engineering controls to keep airborne levels below recommended exposure limits. If user operations generate dust, fume or mist, use ventilation to keep exposure to airborne contaminants below the exposure limit.
**Personal Protection:** Splash goggles. Lab coat. Vapor and dust respirator. Gloves.
**Personal Protection in Case of a Large Spill:** Splash goggles. Full suit. Vapor and dust respirator. Boots. Gloves.
**Exposure Limits:** Not available.

**2-Hydroxyethyl cellulose,** 100 cps CAS# 9004-62-0

| | | | | |
|---|---|---|---|---|
| NFPA/HMIS Rating | Health: 2 | Flammability 1 | Reactivity 0 | Personal Protection E |

Toxicity—Repeated or prolonged exposure is not known to aggravate medical condition.

| | | | | |
|---|---|---|---|---|
| Possible Effects | **Carcinogenic:** NA | **Mutagenic:** NA | **Teratogenic:** NA | **Developmental:** NA |

**Engineering Controls:** Use process enclosures, local exhaust ventilation, or other engineering controls to keep airborne levels below recommended exposure limits. If user operations generate dust, fume or mist, use ventilation to keep exposure to airborne contaminants below the exposure limit.
**Personal Protection:** Splash goggles. Lab coat. Dust respirator. Gloves.
**Personal Protection in Case of a Large Spill:** Splash goggles. Full suit. Dust respirator. Boots. Gloves.
**Exposure Limits:** Not available.

**Table 6–3** Typical Chemicals of Flow-Back Fluid and Health Effects (cont'd)

**Alumina, Aluminum oxide** CAS#1344-28-1

| | | | | |
|---|---|---|---|---|
| NFPA/HMIS Rating | Health: 3 | Flammability 1 | Reactivity 0 | Personal Protection E |

Toxicity—Repeated or prolonged exposure is not known to aggravate medical condition.

| | | | | |
|---|---|---|---|---|
| Possible Effects | **Carcinogenic:** A4 (Not classifiable for human or animal.) | **Mutagenic:** NA | **Teratogenic:** Classified None for human. | **Developmental:** NA |

**Engineering Controls:** Use process enclosures, local exhaust ventilation, or other engineering controls to keep airborne levels below recommended exposure limits. If user operations generate dust, fume or mist, use ventilation to keep exposure to airborne contaminants below the exposure limit.
**Personal Protection:** Splash goggles. Lab coat. Dust respirator. Gloves.
**Personal Protection in Case of a Large Spill:** Splash goggles. Full suit. Vapor and dust respirator. Boots. Gloves.
**Exposure Limits:** TWA: 10 (mg/m3) from ACGIH (TLV) [United States] Inhalation Total. TWA: 10 (mg/m3) [Canada] Inhalation Total. TWA: 5 (mg/m3) from OSHA (PEL) [United States] Inhalation Respirable. TWA: 15 from OSHA (PEL) [United States] Inhalation Total. TWA: 10 [United Kingdom (UK)] Inhalation Total. TWA: 4 [United Kingdom (UK)] Inhalation Respirable.

**Ammonium bisulfate** CAS#7803-63-6

| | | | | |
|---|---|---|---|---|
| NFPA/HMIS Rating | Health: 3 | Flammability 0 | Reactivity 0 | Personal Protection J |

Toxicity—The substance is toxic to lungs, mucous membranes. Repeated or prolonged exposure to the substance can produce target organs damage. Repeated exposure of the eyes to a low level of dust can produce eye irritation. Repeated skin exposure can produce local skin destruction, or dermatitis.

Repeated inhalation of dust can produce varying degree of respiratory irritation or lung damage. Repeated or prolonged inhalation of dust may lead to chronic respiratory irritation.

| | | | | |
|---|---|---|---|---|
| Possible Effects | **Carcinogenic:** NA | **Mutagenic:** NA | **Teratogenic:** NA | **Developmental:** NA |

**Table 6–3** Typical Chemicals of Flow-Back Fluid and Health Effects (cont'd)

**Engineering Controls:** Use process enclosures, local exhaust ventilation, or other engineering controls to keep airborne levels below recommended exposure limits. If user operations generate dust, fume or mist, use ventilation to keep exposure to airborne contaminants below the exposure limit.
**Personal Protection:** Splash goggles. Synthetic apron. Vapor and dust respirator. Gloves.
**Personal Protection in Case of a Large Spill:** Splash goggles. Full suit. Vapor and dust respirator. Boots. Gloves.
**Exposure Limits:** Not available.

**Ammonium chloride** CAS#12125-02-9

| | | | | |
|---|---|---|---|---|
| NFPA/HMIS Rating | Health: 2 | Flammability 0 | Reactivity 0 | Personal Protection E |

Toxicity—Repeated or prolonged exposure is not known to aggravate medical condition.

| | | | | |
|---|---|---|---|---|
| Possible Effects | **Carcinogenic:** NA | **Mutagenic:** NA | **Teratogenic:** NA | **Developmental:** NA |

**Engineering Controls:** Use process enclosures, local exhaust ventilation, or other engineering controls to keep airborne levels below recommended exposure limits. If user operations generate dust, fume or mist, use ventilation to keep exposure to airborne contaminants below the exposure limit.
**Personal Protection:** Splash goggles. Lab coat. Dust respirator. Gloves.
**Personal Protection in Case of a Large Spill:** Splash goggles. Full suit. Dust respirator. Boots. Gloves. A self contained breathing apparatus should be used to avoid inhalation of the product. Suggested protective clothing might not be sufficient; consult a specialist BEFORE handling this product.
**Exposure Limits:** TWA: 10 STEL: 20 (mg/m3) from ACGIH (TLV) [United States] Inhalation TWA: 10 STEL: 20 (mg/m3) [United Kingdom (UK)] Inhalation TWA: 10 STEL: 20 (mg/m3) from NIOSH [United States] Inhalation TWA: 10 STEL: 20 (mg/m3) from OSHA (PEL) [United States]Consult local authorities for acceptable exposure limits.

**Ammonium persulfate** CAS#7727-54-0

| | | | | |
|---|---|---|---|---|
| NFPA/HMIS Rating | Health: 2 | Flammability 1 | Reactivity 3 | Personal Protection E |

Toxicity—The substance may be toxic to upper respiratory tract. Repeated or prolonged exposure to the substance can produce target organs damage.

| | | | | |
|---|---|---|---|---|
| Possible Effects | **Carcinogenic:** NA | **Mutagenic:** NA | **Teratogenic:** NA | **Developmental:** NA |

**Table 6–3** Typical Chemicals of Flow-Back Fluid and Health Effects (cont'd)

**Engineering Controls:** Use process enclosures, local exhaust ventilation, or other engineering controls to keep airborne levels below recommended exposure limits. If user operations generate dust, fume or mist, use ventilation to keep exposure to airborne contaminants below the exposure limit.
**Personal Protection:** Splash goggles. Lab coat. Dust respirator. Gloves.
**Personal Protection in Case of a Large Spill:** Splash goggles. Full suit. Dust respirator. Boots. Gloves.
**Exposure Limits:** Not available.

**Polyacrylamide** CAS#9003-05-8

| | | | | |
|---|---|---|---|---|
| NFPA/HMIS Rating | Health: 2 | Flammability 1 | Reactivity 2 | Personal Protection J |

Toxicity—Repeated or prolonged exposure to the substance can produce target organs damage. Repeated exposure to a highly toxic material may produce general deterioration of health by an accumulation in one or many human organs.

| | | | | |
|---|---|---|---|---|
| Possible Effects | **Carcinogenic:** Classified + (Proven.) by OSHA + (Proven.) by NIOSH. Classified A3 (Proven for animal.) by ACGIH. Classified 2A (Probable for human.) by IARC, 2 (Some evidence.) by NTP, 2 (Suspected for human.) by European Union. | **Mutagenic:** Mutagenic for mammalian somatic cells. Mutagenic for bacteria and/or yeast. | **Teratogenic:** NA | **Developmental:** Classified Reproductive system/toxin/ male [POSSIBLE]. The substance may be toxic to peripheral nervous system, central nervous system (CNS). |

**Engineering Controls:** Use process enclosures, local exhaust ventilation, or other engineering controls to keep airborne levels below recommended exposure limits. If user operations generate dust, fume or mist, use ventilation to keep exposure to airborne contaminants below the exposure limit.

**Table 6–3** Typical Chemicals of Flow-Back Fluid and Health Effects (cont'd)

**Personal Protection:** Splash goggles. Lab coat. Dust respirator. Gloves.
**Personal Protection in Case of a Large Spill:** Splash goggles. Full suit. Dust respirator. Boots. Gloves.
**Exposure Limits:** TWA: 0.03 (mg/m3) [Australia] Inhalation TWA: 0.3 (mg/m3) from OSHA (PEL) [United States] Inhalation TWA: 0.03 (mg/m3) from NIOSH Inhalation TWA: 0.03 (mg/m3) from NIOSH SKIN TWA: 0.3 (mg/m3) [United Kingdom (UK)] Inhalation TWA: 0.03 (mg/m3) from ACGIH (TLV) [United States] [1999] Inhalation Consult local authorities for acceptable exposure limits.

**Benzene** CAS#71-43-2

| | | | | |
|---|---|---|---|---|
| NFPA/HMIS Rating | Health: 2 | Flammability 3 | Reactivity 0 | Personal Protection H |

Toxicity—The substance may be toxic to liver and urinary System. Repeated or prolonged exposure to the substance can produce target organs damage.

| | | | | |
|---|---|---|---|---|
| Possible Effects | **Carcinogenic:** Classified A1 (Confirmed for human.) by ACGIH, 1 (Proven for human.) by IARC. | **Mutagenic:** Classified POSSIBLE for human. Mutagenic for mammalian somatic cells. Mutagenic for bacteria and/or yeast. | **Teratogenic:** Classified None for human. | **Developmental:** Classified Reproductive system/toxin/ female [POSSIBLE]. The substance is toxic to blood, bone marrow, central nervous system (CNS). |

**Engineering Controls:** Provide exhaust ventilation or other engineering controls to keep the airborne concentrations of vapors below their respective threshold limit value. Ensure that eyewash stations and safety showers are proximal to the work-station location.
**Personal Protection:** Splash goggles. Lab coat. Vapor respirator. Gloves.
**Personal Protection in Case of a Large Spill:** Splash goggles. Full suit. Vapor respirator. Boots. Gloves.
**Exposure Limits:** TWA: 0.5 STEL: 2.5 (ppm) from ACGIH (TLV) [United States] TWA: 1.6 STEL: 8 (mg/m3) from ACGIH (TLV) [United States] TWA: 0.1 STEL: 1 from NIOSH TWA: 1 STEL: 5 (ppm) from OSHA (PEL) [United States] TWA: 10 (ppm) from OSHA (PEL) [United States] TWA: 3 (ppm) [United Kingdom (UK)] TWA: 1.6 (mg/m3) [United Kingdom (UK)] TWA: 1 (ppm) [Canada] TWA: 3.2 (mg/m3) [Canada] TWA: 0.5 (ppm) [Canada]Consult local authorities for acceptable exposure limits.

**Table 6–3** Typical Chemicals of Flow-Back Fluid and Health Effects (cont'd)

**Benzyl chloride** CAS#100-44-7

| | | | | |
|---|---|---|---|---|
| NFPA/HMIS Rating | Health: 3 | Flammability 2 | Reactivity 0 | Personal Protection |

Toxicity The substance is toxic to kidneys, lungs, the nervous system, liver, mucous membranes. Repeated or prolonged exposure to the substance can produce target organs damage. Repeated or prolonged contact with spray mist may produce chronic eye irritation and severe skin irritation. Repeated or prolonged exposure to spray mist may produce respiratory tract irritation leading to frequent attacks of bronchial infection. Repeated exposure to a highly toxic material may produce general deterioration of health by an accumulation in one or many human organs. Repeated or prolonged inhalation of vapors may lead to chronic respiratory irritation.

| Possible Effects | **Carcinogenic:** | **Mutagenic:** | **Teratogenic:** | **Developmental:** |
|---|---|---|---|---|
| | Classified A2 (Suspected for human.) by ACGIH | NA | NA | NA |

**Engineering Controls:** Provide exhaust ventilation or other engineering controls to keep the airborne concentrations of vapors below their respective threshold limit value. Ensure that eyewash stations and safety showers are proximal to the work-station location.
**Personal Protection:** Face shield. Full suit. Vapor respirator. Gloves. Boots.
**Personal Protection in Case of a Large Spill:** Splash goggles. Full suit. Vapor respirator. Boots. Gloves.
**Exposure Limits:** TWA: 1 Consult local authorities for acceptable exposure limits.

**Beryllium sulfate tetrahydrate** CAS#7787-56-6

| | | | | |
|---|---|---|---|---|
| NFPA/HMIS Rating | Health: 2 | Flammability 0 | Reactivity 0 | Personal Protection E |

Toxicity—The substance is toxic to blood, lungs, the reproductive system, upper respiratory tract, eyes. Repeated or prolonged exposure to the substance can produce target organs damage. Repeated exposure to a highly toxic material may produce general deterioration of health by an accumulation in one or many human organs.

| Possible Effects | **Carcinogenic:** | **Mutagenic:** | **Teratogenic:** | **Developmental:** |
|---|---|---|---|---|
| | Classified 2 (Some evidence.) by NTP. | NA | NA | Classified Reproductive system/toxin/ female, Reproductive system/toxin/ male [PROVEN]. |

**Table 6–3** Typical Chemicals of Flow-Back Fluid and Health Effects (cont'd)

**Engineering Controls:** Use process enclosures, local exhaust ventilation, or other engineering controls to keep airborne levels below recommended exposure limits. If user operations generate dust, fume or mist, use ventilation to keep exposure to airborne contaminants below the exposure limit.
**Personal Protection:** Splash goggles. Lab coat. Dust respirator. Gloves.
**Personal Protection in Case of a Large Spill:** Splash goggles. Full suit. Dust respirator. Boots. Gloves.
**Exposure Limits:** TWA: 0.002 STEL: 0.005 CEIL: 0.025 from ACGIH (TLV) [United States] Consult local authorities for acceptable exposure limits.

**Boric acid** CAS#10043-35-3

| | | | | |
|---|---|---|---|---|
| NFPA/HMIS Rating | Health: 2 | Flammability 0 | Reactivity 0 | Personal Protection E |

Toxicity—The substance may be toxic to kidneys, cardiovascular system, central nervous system (CNS). Repeated or prolonged exposure to the substance can produce target organs damage.

| | | | | |
|---|---|---|---|---|
| Possible Effects | **Carcinogenic:** NA | **Mutagenic:** Mutagenic for bacteria and/or yeast. | **Teratogenic:** NA | **Developmental:** Classified Reproductive system/toxin/ female, Reproductive system/toxin/ male [POSSIBLE]. |

**Engineering Controls:** Use process enclosures, local exhaust ventilation, or other engineering controls to keep airborne levels below recommended exposure limits. If user operations generate dust, fume or mist, use ventilation to keep exposure to airborne contaminants below the exposure limit.
**Personal Protection:** Splash goggles. Lab coat. Dust respirator. Gloves.
**Personal Protection in Case of a Large Spill:** Splash goggles. Full suit. Dust respirator. Boots. Gloves.
**Exposure Limits:** Not available.

**Citric acid** CAS#77-92-9

| | | | | |
|---|---|---|---|---|
| NFPA/HMIS Rating | Health: 2 | Flammability 1 | Reactivity 0 | Personal Protection E |

Toxicity—The substance may be toxic to teeth. Repeated or prolonged exposure to the substance can produce target organs damage. Repeated exposure of the eyes to a low level of dust can produce eye irritation. Repeated skin exposure can produce local skin destruction, or dermatitis. Repeated inhalation of dust can produce varying degree of respiratory irritation or lung damage.

**Table 6–3** Typical Chemicals of Flow-Back Fluid and Health Effects (cont'd)

| | | | | |
|---|---|---|---|---|
| Possible Effects | **Carcinogenic:** NA | **Mutagenic:** NA | **Teratogenic:** NA | **Developmental:** NA |

**Engineering Controls:** Use process enclosures, local exhaust ventilation, or other engineering controls to keep airborne levels below recommended exposure limits. If user operations generate dust, fume or mist, use ventilation to keep exposure to airborne contaminants below the exposure limit.
**Personal Protection:** Safety glasses. Lab coat. Gloves (impervious). Dust respirator. The dust respirator should be used for conditions where exposure has exceeded recommended exposure limits, dust is apparent, and engineering controls(adequate ventilation) are not feasible.
**Personal Protection in Case of a Large Spill:** Splash goggles. Full suit. Dust respirator. Boots. Gloves.
**Exposure Limits:** No exposure guidelines have been established. ACGIH, NIOSH and OSHA have not developed exposure limits for this product. The exposure limits given below are for particulates not otherwise classified: ACGIH: 10 mg/m3 TWA (Total Inhalable fraction); 3 mg/m3 TWA (Respirable fraction) OSHA: 15 mg/m3 TWA (Total dust); 5 mg/m3 TWA (Respirable Fraction)

**Diesel Fuel** CAS#68476-34-6

| | | | | |
|---|---|---|---|---|
| NFPA/HMIS Rating | Health: 2 | Flammability 0 | Reactivity 0 | Personal Protection |

Toxicity—Slightly hazardous in case of skin contact (irritant, sensitizer).

| | | | | |
|---|---|---|---|---|
| Possible Effects | **Carcinogenic:** NA | **Mutagenic:** NA | **Teratogenic:** NA | **Developmental:** NA |

**Engineering Controls:** Use adequate ventilation to keep vapor concentrations of this product below occupational exposure and flammability limits, particularly in confined spaces.
**Personal Protection:** Splash goggles. Lab coat. Vapor respirator. Gloves.
**Personal Protection in Case of a Large Spill:** Splash goggles. Full suit. Dust respirator. Boots. Gloves.
**Exposure Limits:** Refer to OSHA 29 CFR 1910.134, NIOSH Respirator Decision Logic, and the manufacturer for additional guidance on respiratory protection selection. Use a positive pressure, air-supplied respirator if there is a potential for uncontrolled release, exposure levels are not known, in oxygen-deficient atmospheres, or any other circumstance where an air-purifying respirator may not provide adequate protection.

**Ethylbenzene** CAS#100-41-4

| | | | | |
|---|---|---|---|---|
| NFPA/HMIS Rating | Health: 2 | Flammability 3 | Reactivity 0 | Personal Protection H |

**Table 6–3** Typical Chemicals of Flow-Back Fluid and Health Effects (cont'd)

Toxicity—The substance may be toxic to central nervous system (CNS). Repeated or prolonged exposure to the substance can produce target organs damage.

| | | | | |
|---|---|---|---|---|
| Possible Effects | **Carcinogenic:** Classified 2B (Possible for human.) by IARC. | **Mutagenic:** Mutagenic for mammalian somatic cells. Mutagenic for bacteria and/or yeast. | **Teratogenic:** NA | **Developmental:** NA |

**Engineering Controls:** Provide exhaust ventilation or other engineering controls to keep the airborne concentrations of vapors below their respective threshold limit value. Ensure that eyewash stations and safety showers are proximal to the work-station location.
**Personal Protection:** Splash goggles. Lab coat. Vapor respirator. Gloves.
**Personal Protection in Case of a Large Spill:** Splash goggles. Full suit. Vapor respirator. Boots. Gloves.
**Exposure Limits:** TWA: 100 STEL: 125 (ppm) from OSHA (PEL) [United States] TWA: 435 STEL: 545 from OSHA (PEL) [United States] TWA: 435 STEL: 545 (mg/m3) from NIOSH [United States] TWA: 100 STEL: 125 (ppm) from NIOSH [United States] TWA: 100 STEL: 125 (ppm) from ACGIH (TLV) [United States] TWA: 100 STEL: 125 (ppm) [United Kingdom (UK)] TWA: 100 STEL: 125 (ppm) [Belgium] TWA: 100 STEL: 125 (ppm) [Finland] TWA: 50 (ppm) [Norway] Consult local authorities for acceptable exposure limits.

**Ethylene glycol** CAS#107-21-1

| | | | | |
|---|---|---|---|---|
| NFPA/HMIS Rating | Health: 1 | Flammability 1 | Reactivity 0 | Personal Protection C |

Toxicity—The substance may be toxic to kidneys, liver, central nervous system (CNS). Repeated or prolonged exposure to the substance can produce target organs damage. Repeated exposure to a highly toxic material may produce general deterioration of health by an accumulation in one or many human organs.

| | | | | |
|---|---|---|---|---|
| Possible Effects | **Carcinogenic:** A4 (Not classifiable for human or animal.) by ACGIH. | **Mutagenic:** Mutagenic for mammalian somatic cells. Non-mutagenic for bacteria and/or yeast. | **Teratogenic:** NA | **Developmental:** NA |

**Table 6–3** Typical Chemicals of Flow-Back Fluid and Health Effects (cont'd)

**Engineering Controls:** Provide exhaust ventilation or other engineering controls to keep the airborne concentrations of vapors below their respective threshold limit value. Ensure that eyewash stations and safety showers are proximal to the work-station location.
**Personal Protection:** Safety glasses. Synthetic apron. Gloves (impervious). For most conditions, no respiratory protection should be needed. However, if material is heated or sprayed and if atmospheric levels exceed exposure guidelines, use an approved vapor (air purifying) respirator.
**Personal Protection in Case of a Large Spill:** Splash goggles. Full suit. Boots. Gloves. Suggested protective clothing might not be sufficient; consult a specialist BEFORE handling this product.
**Exposure Limits:** STEL: 120 (mg/m3) [Australia] TWA: 100 (mg/m3) from ACGIH (TLV) [United States] CEIL: 125 (mg/m3) from OSHA (PEL) [United States] CEIL: 50 (ppm) from OSHA (PEL) [United States] TWA: 52 STEL: 104 (mg/m3) [United Kingdom (UK)] Inhalation TWA: 10 (mg/m3) [United Kingdom (UK)] SKIN3 Consult local authorities for acceptable exposure limits.

**Glutaraldehyde Solution,** 50% CAS# 111-30-8 (**Water,** 50% CAS#7732-18-5)

| | | | | |
|---|---|---|---|---|
| NFPA/HMIS Rating | Health: 2 | Flammability 0 | Reactivity 0 | Personal Protection H |

Toxicity—The substance may be toxic to blood, the reproductive system, liver, mucous membranes, spleen, central nervous system (CNS), Urinary System. Repeated or prolonged exposure to the substance can produce target organs damage. Repeated or prolonged contact with spray mist may produce chronic eye irritation and severe skin irritation. Repeated or prolonged exposure to spray mist may produce respiratory tract irritation leading to frequent attacks of bronchial infection. Repeated exposure to a highly toxic material may produce general deterioration of health by an accumulation in one or many human organs.

| | | | | |
|---|---|---|---|---|
| Possible Effects | **Carcinogenic:** Classified A4 (Not classifiable for human or animal.) by ACGIH [Glutaraldehyde]. | **Mutagenic:** Mutagenic for mammalian somatic cells. Mutagenic for bacteria and/or yeast. | **Teratogenic:** NA | **Developmental:** Classified Reproductive system/toxin/female, Reproductive system/toxin/male [SUSPECTED] |

**Engineering Controls:** Provide exhaust ventilation or other engineering controls to keep the airborne concentrations of vapors below their respective threshold limit value. Ensure that eyewash stations and safety showers are proximal to the work-station location.

**Table 6–3** Typical Chemicals of Flow-Back Fluid and Health Effects (cont'd)

**Personal Protection:** Splash goggles. Lab coat. Vapor respirator. Gloves.
**Personal Protection in Case of a Large Spill:** Splash goggles. Full suit. Vapor respirator. Boots. Gloves.
**Exposure Limits:** Glutaraldehyde TWA: 0.2 (ppm) [Australia] TWA: 0.82 (mg/m3) [Australia] TWA: 0.25 CEIL: 0.2 (ppm) from NIOSH CEIL: 0.2 (ppm) from OSHA (PEL) [United States] TWA: 0.05 STEL: 0.05 (ppm) [United Kingdom (UK)] Consult local authorities for acceptable exposure limits.

**Guar Gum** CAS#9000-30-0

| | | | | |
|---|---|---|---|---|
| NFPA/HMIS Rating | Health: 1 | Flammability 1 | Reactivity 0 | Personal Protection E |

Toxicity—Repeated or prolonged exposure is not known to aggravate medical condition.

| | | | | |
|---|---|---|---|---|
| Possible Effects | **Carcinogenic:** NA | **Mutagenic:** NA | **Teratogenic:** NA | **Developmental:** NA |

**Engineering Controls:** Use process enclosures, local exhaust ventilation, or other engineering controls to keep airborne levels below recommended exposure limits. If user operations generate dust, fume or mist, use ventilation to keep exposure to airborne contaminants below the exposure limit.
**Personal Protection:** Safety glasses. Lab coat. Dust respirator. Gloves.
**Personal Protection in Case of a Large Spill:** Splash goggles. Full suit. Dust respirator. Boots. Gloves.
**Exposure Limits:** Not available.

**Hydrous Silicate of Alumina** CAS# 1302-78-9, **Crystalline Silica in the form of Quartz** CAS# 14808-60-7

| | | | | |
|---|---|---|---|---|
| NFPA/HMIS Rating | Health: 1 | Flammability 0 | Reactivity 0 | Personal Protection E |

Toxicity—The substance may be toxic to lungs, upper respiratory tract. Repeated or prolonged exposure to the substance can produce target organs damage.

| | | | | |
|---|---|---|---|---|
| Possible Effects | **Carcinogenic:** Classified 1 (Proven for human.) by IARC. Classified A2 (Suspected for human.) by ACGIH | **Mutagenic:** NA | **Teratogenic:** NA | **Developmental:** NA |

**Table 6–3** Typical Chemicals of Flow-Back Fluid and Health Effects (cont'd)

**Engineering Controls:** Use process enclosures, local exhaust ventilation, or other engineering controls to keep airborne levels below recommended exposure limits. If user operations generate dust, fume or mist, use ventilation to keep exposure to airborne contaminants below the exposure limit.
**Personal Protection:** Safety glasses. Lab coat. Dust respirator. Gloves.
**Personal Protection in Case of a Large Spill:** Splash goggles. Full suit. Dust respirator. Boots. Gloves.
**Exposure Limits:** TWA: 0.05 (mg/m3) from ACGIH (TLV) [United States] Respirable [Quartz or Crystalline Silica] TWA: 0.05 (mg/m3) from NIOSH Respirable. [Quartz or Crystalline Silica]Consult local authorities for acceptable exposure limits

**Isopropyl alcohol** CAS#67-63-0

| | | | | |
|---|---|---|---|---|
| NFPA/HMIS Rating | Health: 2 | Flammability 3 | Reactivity 0 | Personal Protection H |

Toxicity—The substance may be toxic to kidneys, liver, skin, central nervous system (CNS). Repeated or prolonged exposure to the substance can produce target organs damage

| | | | | |
|---|---|---|---|---|
| Possible Effects | **Carcinogenic:** A4 (Not classifiable for human or animal.) by ACGIH, 3 (Not classifiable for human.) by IARC. | **Mutagenic:** NA | **Teratogenic:** NA | **Developmental:** Classified Reproductive system/toxin/ female, Development toxin [POSSIBLE]. |

**Engineering Controls:** Provide exhaust ventilation or other engineering controls to keep the airborne concentrations of vapors below their respective threshold limit value. Ensure that eyewash stations and safety showers are proximal to the work-station location.
**Personal Protection:** Splash goggles. Lab coat. Vapor respirator. Gloves.
**Personal Protection in Case of a Large Spill:** Splash goggles. Full suit. Vapor respirator. Boots. Gloves.
**Exposure Limits:** TWA: 983 STEL: 1230 (mg/m3) [Australia] TWA: 200 STEL: 400 (ppm) from ACGIH (TLV) [United States] [1999] TWA: 980 STEL: 1225 (mg/m3) from NIOSH TWA: 400 STEL: 500 (ppm) from NIOSH TWA: 400 STEL: 500 (ppm) [United Kingdom (UK)] TWA: 999 STEL: 1259 (mg/m3) [United Kingdom (UK)] TWA: 400 STEL: 500 (ppm) from OSHA (PEL) [United States] TWA: 980 STEL: 1225 (mg/m3) from OSHA (PEL) [United States]Consult local authorities for acceptable exposure limits.

**Table 6–3** Typical Chemicals of Flow-Back Fluid and Health Effects (cont'd)

**Mineral oil** CAS#8042-47-5 (previously 8012-95-1)

| | | | | |
|---|---|---|---|---|
| NFPA/HMIS Rating | Health: 1 | Flammability 1 | Reactivity 0 | Personal Protection A |

Toxicity—Repeated or prolonged exposure is not known to aggravate medical condition.

| | | | | |
|---|---|---|---|---|
| Possible Effects | **Carcinogenic:** 3 (Not classifiable for human.) by IARC. | **Mutagenic:** NA | **Teratogenic:** NA | **Developmental:** NA |

**Engineering Controls:** Provide exhaust ventilation or other engineering controls to keep the airborne concentrations of vapors below their respective threshold limit value. Ensure that eyewash stations and safety showers are proximal to the work-station location.
**Personal Protection:** Safety glasses. Lab coat.
**Personal Protection in Case of a Large Spill:** Splash goggles. Full suit. Boots. Gloves.
**Exposure Limits:** TWA: 5 STEL: 10 (mg/m3) as oil mist Consult local authorities for acceptable exposure limits

**N,N-Dimethylformamide** CAS# 68-12-2

| | | | | |
|---|---|---|---|---|
| NFPA/HMIS Rating | Health: 2 | Flammability 2 | Reactivity 0 | Personal Protection H |

Toxicity—The substance is toxic to kidneys, liver, central nervous system (CNS). The substance may be toxic to blood, the nervous system. Repeated or prolonged exposure to the substance can produce target organs damage.

| | | | | |
|---|---|---|---|---|
| Possible Effects | **Carcinogenic:** 3 (Not classifiable for human.) by IARC. | **Mutagenic:** Mutagenic for mammalian somatic cells. Mutagenic for bacteria and/or yeast. | **Teratogenic:** Classified POSSIBLE for human. | **Developmental:** **Classified** Reproductive system/toxin/ female, Reproductive system/toxin/ male [POSSIBLE]. |

**Engineering Controls:** Provide exhaust ventilation or other engineering controls to keep the airborne concentrations of vapors below their respective threshold limit value. Ensure that eyewash stations and safety showers are proximal to the work-station location.

**Table 6–3** Typical Chemicals of Flow-Back Fluid and Health Effects (cont'd)

**Personal Protection:** Splash goggles. Lab coat. Vapor respirator. Gloves.
**Personal Protection in Case of a Large Spill:** Splash goggles. Full suit. Vapor respirator. Boots. Gloves.
**Exposure Limits:** TWA: 10 (ppm) from ACGIH (TLV) [United States] TWA: 30 (mg/m3) from ACGIH (TLV) [United States] Consult local authorities for acceptable exposure limits.

**Potassium carbonate, anhydrous** CAS#584-08-7

| | | | | |
|---|---|---|---|---|
| NFPA/HMIS Rating | Health: 2 | Flammability 0 | Reactivity 1 | Personal Protection E |

Toxicity—The substance is toxic to mucous membranes. The substance may be toxic to skin, eyes. Repeated or prolonged exposure to the substance can produce target organs damage.

| | | | | |
|---|---|---|---|---|
| Possible Effects | **Carcinogenic:** NA | **Mutagenic:** NA | **Teratogenic:** NA | **Developmental:** NA |

**Engineering Controls:** Use process enclosures, local exhaust ventilation, or other engineering controls to keep airborne levels below recommended exposure limits. If user operations generate dust, fume or mist, use ventilation to keep exposure to airborne contaminants below the exposure limit.
**Personal Protection:** Splash goggles. Lab coat. Dust respirator. Gloves.
**Personal Protection in Case of a Large Spill:** Splash goggles. Full suit. Dust respirator. Boots. Gloves.
**Exposure Limits:** Not available.

**Potassium chloride** CAS#7447-40-7

| | | | | |
|---|---|---|---|---|
| NFPA/HMIS Rating | Health: 1 | Flammability 0 | Reactivity 0 | Personal Protection E |

Toxicity—The substance may be toxic to blood, cardiovascular system. Repeated or prolonged exposure to the substance can produce target organs damage.

| | | | | |
|---|---|---|---|---|
| Possible Effects | **Carcinogenic:** NA | **Mutagenic:** Mutagenic for mammalian somatic cells. Mutagenic for bacteria and/or yeast. | **Teratogenic:** NA | **Developmental:** NA |

**Engineering Controls:** Use process enclosures, local exhaust ventilation, or other engineering controls to keep airborne levels below recommended exposure limits. If user operations generate dust, fume or mist, use ventilation to keep exposure to airborne contaminants below the exposure limit.

**Table 6–3** Typical Chemicals of Flow-Back Fluid and Health Effects (cont'd)

**Personal Protection:** Safety glasses. Lab coat. Dust respirator. Gloves.
**Personal Protection in Case of a Large Spill:** Splash goggles. Full suit. Dust respirator. Boots. Gloves.
**Exposure Limits:** Not available.

**Propylene glycol** CAS#57-55-6

| | | | | |
|---|---|---|---|---|
| NFPA/HMIS Rating | Health: 2 | Flammability 1 | Reactivity 0 | Personal Protection H |

Toxicity—The substance may be toxic to central nervous system (CNS). Repeated or prolonged exposure to the substance can produce target organs damage.

| | | | | |
|---|---|---|---|---|
| Possible Effects | **Carcinogenic:** NA | **Mutagenic:** NA | **Teratogenic:** NA | **Developmental:** NA |

**Engineering Controls:** Provide exhaust ventilation or other engineering controls to keep the airborne concentrations of vapors below their respective threshold limit value. Ensure that eyewash stations and safety showers are proximal to the work-station location.
**Personal Protection:** Splash goggles. Lab coat. Vapor respirator. Gloves.
**Personal Protection in Case of a Large Spill:** Splash goggles. Full suit. Vapor respirator. Boots. Gloves.
**Exposure Limits:** TWA: 10 (mg/m3) from AIHA Consult local authorities for acceptable exposure limits.

**Silica/Sand (Crystalline Silica (quartz) SiO2)** 99.0–99.9% CAS# 14808-60-7

| | | | | |
|---|---|---|---|---|
| NFPA/HMIS Rating | Health: 1 | Flammability 0 | Reactivity 0 | Personal Protection E |

Toxicity—Repeated or prolonged exposure of individuals with lung disease (e.g., bronchitis, emphysema, chronic obstructive pulmonary disease) can be aggravated by exposure.

| | | | | |
|---|---|---|---|---|
| Possible Effects | **Carcinogenic:** NA | **Mutagenic:** NA | **Teratogenic:** NA | **Developmental:** NA |

**Engineering Controls:** Use process enclosures, local exhaust ventilation, or other engineering controls to keep airborne levels below recommended exposure limits. If user operations generate dust, fume or mist, use ventilation to keep exposure to airborne contaminants below the exposure limit.
**Personal Protection:** Safety glasses. Lab coat. Dust respirator. Gloves.
**Personal Protection in Case of a Large Spill:** Splash goggles. Full suit. Dust respirator. Boots. Gloves.

**Table 6–3** Typical Chemicals of Flow-Back Fluid and Health Effects (cont'd)

**Exposure Limits:** TLV = 0.025 mg/m3 as respirable quartz particles median size less than 4 um.

**Sodium borate** CAS#1303-96-4

| | | | | |
|---|---|---|---|---|
| NFPA/HMIS Rating | Health: 2 | Flammability 0 | Reactivity 0 | Personal Protection E |

Toxicity—Repeated or prolonged exposure is not known to aggravate medical condition.

| | | | | |
|---|---|---|---|---|
| Possible Effects | **Carcinogenic:** NA | **Mutagenic:** NA | **Teratogenic:** NA | **Developmental:** NA |

**Engineering Controls:** Use process enclosures, local exhaust ventilation, or other engineering controls to keep airborne levels below recommended exposure limits. If user operations generate dust, fume or mist, use ventilation to keep exposure to airborne contaminants below the exposure limit.
**Personal Protection:** Splash goggles. Lab coat. Dust respirator. Gloves.
**Personal Protection in Case of a Large Spill:** Splash goggles. Full suit. Dust respirator. Boots. Gloves.
**Exposure Limits:** TWA: 0.31 (ppm) TWA: 5 (mg/m3) from ACGIH [1995] Consult local authorities for acceptable exposure limits.

**Sodium carbonate** CAS#497-19-8

| | | | | |
|---|---|---|---|---|
| NFPA/HMIS Rating | Health: 2 | Flammability 0 | Reactivity1 | Personal Protection E |

Toxicity—The substance may be toxic to upper respiratory tract, skin, eyes. Repeated or prolonged exposure to the substance can produce target organs damage.

| | | | | |
|---|---|---|---|---|
| Possible Effects | **Carcinogenic:** NA | **Carcinogenic:** NA | **Teratogenic:** NA | **Developmental:** NA |

**Engineering Controls:** Use process enclosures, local exhaust ventilation, or other engineering controls to keep airborne levels below recommended exposure limits. If user operations generate dust, fume or mist, use ventilation to keep exposure to airborne contaminants below the exposure limit.
**Personal Protection:** Splash goggles. Lab coat. Dust respirator. Gloves.
**Personal Protection in Case of a Large Spill:** Splash goggles. Full suit. Dust respirator. Boots. Gloves.
**Exposure Limits:** TWA: 0.31 (ppm) TWA: 5 (mg/m3) from ACGIH [1995] Consult local authorities for acceptable exposure limits.

**Table 6–3** Typical Chemicals of Flow-Back Fluid and Health Effects (cont'd)

**Toluene** CAS#108-88-3

| | | | | |
|---|---|---|---|---|
| NFPA/HMIS Rating | Health: 2 | Flammability 3 | Reactivity 0 | Personal Protection H |

Toxicity—The substance may be toxic to blood, kidneys, the nervous system, liver, brain, central nervous system (CNS). Repeated or prolonged exposure to the substance can produce target organs damage.

| | | | | |
|---|---|---|---|---|
| Possible Effects | **Carcinogenic:** A4 (Not classifiable for human or animal.) by ACGIH, 3 (Not classifiable for human.) by IARC. | **Mutagenic:** NA | **Teratogenic:** NA | **Developmental:** NA |

**Engineering Controls:** Provide exhaust ventilation or other engineering controls to keep the airborne concentrations of vapors below their respective threshold limit value. Ensure that eyewash stations and safety showers are proximal to the work-station location.
**Personal Protection:** Splash goggles. Lab coat. Vapor respirator. Gloves.
**Personal Protection in Case of a Large Spill:** Splash goggles. Full suit. Vapor respirator. Boots. Gloves.
**Exposure Limits:** TWA: 200 STEL: 500 CEIL: 300 (ppm) from OSHA (PEL) [United States] TWA: 50 (ppm) from ACGIH (TLV) [United States] SKIN TWA: 100 STEL: 150 from NIOSH [United States] TWA: 375 STEL: 560 (mg/m3) from NIOSH [United States] Consult local authorities for acceptable exposure limits

**Xylenes** CAS#1330-20-7

| | | | | |
|---|---|---|---|---|
| NFPA/HMIS Rating | Health: 2 | Flammability 3 | Reactivity 0 | Personal Protection H |

Toxicity—The substance may be toxic to blood, kidneys, liver, mucous membranes, bone marrow, central nervous system (CNS). Repeated or prolonged exposure to the substance can produce target organs damage.

| | | | | |
|---|---|---|---|---|
| Possible Effects | **Carcinogenic:** 3 (Not classifiable for human.) by IARC. | **Mutagenic:** NA | **Teratogenic:** NA | **Developmental:** NA |

**Table 6–3** Typical Chemicals of Flow-Back Fluid and Health Effects (cont'd)

**Engineering Controls:** Provide exhaust ventilation or other engineering controls to keep the airborne concentrations of vapors below their respective threshold limit value. Ensure that eyewash stations and safety showers are proximal to the work-station location.
**Personal Protection:** Splash goggles. Lab coat. Vapor respirator. Gloves.
**Personal Protection in Case of a Large Spill:** Splash goggles. Full suit. Vapor respirator. Boots. Gloves.
**Exposure Limits:** TWA: 100 (ppm) [Canada] TWA: 435 (mg/m3) [Canada] TWA: 434 STEL: 651 (mg/m3) from ACGIH (TLV) [United States] TWA: 100 STEL: 150 (ppm) from ACGIH (TLV) [United States] Consult local authorities for acceptable exposure limits.

# APPENDIX A

# Viscosity

The varied behavior of viscosity is of such great importance to the technology of hydraulic fracturing that an extended explanation of its various conditions is presented.

The environmental properties of each formation (rock consolidation, temperature, composition of pore fluids, etc.) are so different that each one requires preparation of a frac-fluid specifically tailored for the well and formation to be fractured. The most important considerations for development of the frac-fluid are creation of the fracture and its width and length, which are functions of the apparent viscosity as a function of time and fluid efficiency (fluid loss to the formation matrix). Fluids with high values of viscosity (and the resulting elevated pressures that are required for pumping into the formation) promote the formation of wide fractures of greater length; but the capacity of the pumps and cost of additives required to prepare the frac-fluid must also be considered. In addition, the proppant suspension (or proppant-carrying ability), temperature stability of the frac-fluid, possible formation damage caused by injection of the fracture fluid, and loss of pressure due to fluid friction in the well pipes (well tubulars) are other important considerations.

The terms of the fluid models that express viscosity are presented (for complete understanding of the physical conditions in action during the flow of fluids) in confined mediums (pipes, etc). Stress, $\tau_x$, is the *force per unit area* (or pressure), exerted in the x-direction, causing the shearing motion of the fluid represented in a rectangular pipe (see Figure A–1):

$$\tau_x = \frac{F}{A} \quad \left[\frac{g_{force}}{cm^2}; \frac{N(10^{-1})}{m^2}; Pa(10^{-1}); \frac{lbs_{force}}{ft^2}\right] \tag{A.1}$$

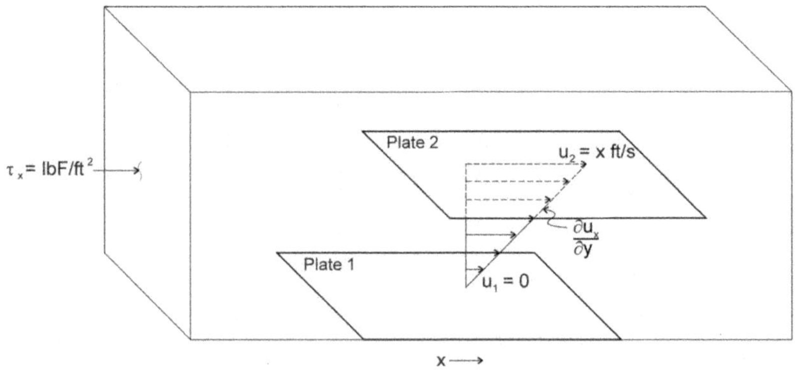


**Figure A–1** *Flow of a Newtonian fluid in a rectangular pipe: Plate 2, in the center, is moving at velocity $u_x$ relative to Plate 1, which is stationary. A Newtonian fluid exhibits a constant shear rate ($du_x/dy$) from $u_{x-1} = 0$ at Plate 1, to $u_{x-2} = x$ ft/s.*

In Figure A–1, the shear stress causes a fluid in a rectangular pipe to move in the x-direction through the pipe. If laminar flow is taking place, the process can be imagined as one in which parallel planes of fluid slide past each other. In other words, the fluid flows through the pipe as segments of parallel plates such as the one in the center. Because of friction between the fluid and the walls on each side of the pipe, the velocity of each succeeding plate (stacked from the center to the bottom) is less than the other as each approaches the bottom of the pipe, as shown in Figure A–1. The bottom plate is stationary while the upper plate moves with velocity $u_x$ in the x-direction and the fluid adopts a linear velocity profile, or change of velocity ($u_x$) with respect to y (the distance between the plates)

$$\left(\frac{\delta u_x}{\delta y}\right)$$

due to friction along the bottom surface. The velocity profile (or velocity gradient) is the *shear rate;* in other words, the difference in velocity of the plates with respect to the distance from the center plate (the plate with the greatest velocity) to the bottom plate, whose velocity is zero (see Figure A–1):

$$\frac{\delta u_x}{\delta y} = \frac{u_1 - u_2}{y} \quad \left[\frac{cm/s}{cm}; \frac{ft/s}{ft} = \frac{1}{s}\right] \tag{A.2}$$

The apparent viscosity ($\mu_a$) is defined as the ratio of shear stress to shear rate:

$$\mu_a = \frac{\tau_x}{\left(\frac{\delta u_x}{\delta_y}\right)} \tag{A.3}$$

where

$u_x$ = laminar fluid velocity in the x-direction

y = distance from the center to the bottom of the pipe (Figure A–2)

$\mu_a$ = apparent viscosity at a specific time and temperature

$\tau_x$ = shear stress in the x-direction

Various types of fluids are described by their fluid-flow behavior as: Newtonian, Bingham, Power-law Dilatant, Power-law Pseudoplastic, Rheopectic, Thixotropic, etc.

**Newtonian Fluids:** these are the simplest because they exhibit constant viscosity at all shear rates and fluid velocities (see Curve 1, Figure A–2). For a Newtonian fluid, the apparent viscosity is equal to the absolute viscosity on any point on the line, thus:

$$\mu_a = \mu = \frac{\tau_x}{\left(\frac{\delta u_x}{\delta y}\right)} \tag{A.4}$$

The SI units of viscosity are [kg/m-s = Pa-s (Pascal-second)]. Generally, however, the Poise and centipoise (cP) are used, and the conversions of one to the other of various units that are encountered are:

Poise = 100cP = 0.1 N-s/m$^2$ = 0.1 Pa-s = 0.1 kg/m-s = 0.0672 lbM/ft-s = 0.002089 lbF/ft-s.

The terms *apparent viscosity* ($\mu_a$), *absolute viscosity* ($\mu$) and *shear viscosity* are synonymous with the viscosity as defined in Eq. (A.4).

Examples of Newtonian fluids are: all gases and pure low-molecular weight liquids.

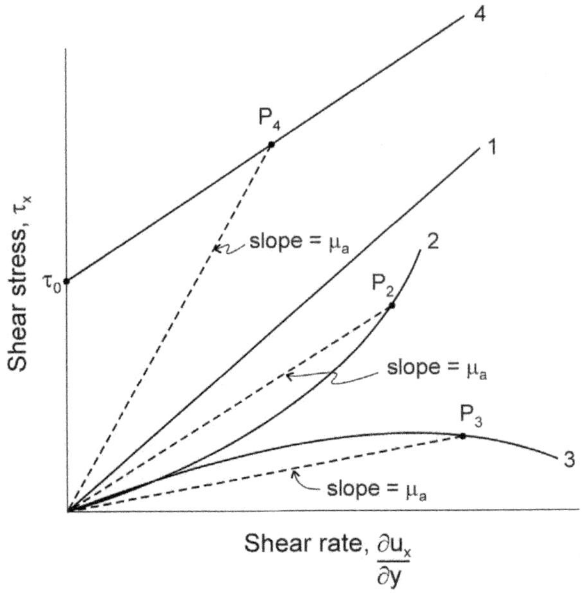


**Figure A–2** *Curve 1: Newtonian fluid whose slope, at any point on the line, is equal to the viscosity (μ). Curve 2: Dilatent (shear thickening) fluid whose viscosity increases as the shear rate increases. Curve 3: Pseudoplastic fluid (shear thinning) whose viscosity decreases as the shear rate increases (most frac-fluids). Curve 4: Bingham fluid that requires an initial, finite value of shear stress ($\tau_0$) before fluid flow will occur, thus they are sometimes referred to as yield stress fluids. The apparent viscosity, at any point $P_4$, is defined as the slope of the line drawn from point $P_4$ to the origin.*

**Non-Newtonian Fluids:** fluids exhibit viscosities that change with respect to the rate of shear (see Curves 2 and 3, Figure A–2). Non-Newtonian fluids may be represented by an equation such as:

$$\tau_w = \left[\beta\left(\frac{\delta u_x}{\delta y}\right)\right]_f \tag{A.5}$$

where

$\tau_w$ = wall shear stress (equal to $\tau_x$)

$\left[\beta\left(\frac{\delta u_x}{\delta y}\right)\right]_f$ = function of the velocity gradient at any point under consideration

$u_x$ = velocity in the x-direction

y = center to the wall distance (distance from Plate 2 to Plate 1, Figure A–1)

Equating Eqs. (A.4 and A.5) yields:

$$\mu_a = \frac{[\beta(\delta u_x / \delta y)]_f}{\delta u_x / \delta y} \tag{A.6}$$

From Eq. (A.6), if any point ($P_2$, $P_3$, $P_4$) on a fluid-flow curve of Figure A–2 is joined to the origin (dashed lines), the slope of the dashed line is equal to the apparent viscosity at that specific shear rate.

## A.1 Power-Law Fluids

There are many types of non-Newtonian fluids; those that exhibit viscosities that change (becoming higher or lower) with respect to flow rate (hence the shear rate or velocity gradient of Figure A–2) and are represented mathematically by a *Power-Law* model:

$$\tau_x = K\left(\frac{\delta u_x}{\delta y}\right)^n \tag{A.7}$$

where

K = consistency index ($\tau_x$ intercept at a shear-rate equal to 1.0 $sec^{-1}$)

n = flow behavior index

The apparent viscosity is defined as:

$$u_a = K\left(\frac{\delta u_x}{\delta y}\right)^{n-1} \tag{A.8}$$

**Dilatent Fluids:** Dilatent fluids are characterized by an apparent viscosity that *increases* with respect to the shear rate (see Curve 2, Figure A–2), and

$$\mu_a = K\left(\frac{\delta u_x}{\delta y}\right)^{n-1} ; \quad n > 1 \tag{A.9}$$

Examples of dilatent fluids are: polymer solutions and starch suspensions.

**Pseudoplastic Fluids:** The apparent viscosity of pseudoplastic fluids *decreases* with respect to the shear rate (see Curve 3, Figure A–2). These are generally considered to be the greatest class of non-Newtonian fluids, and they are represented by the following equation:

$$u_a = K\left(\frac{\delta u_x}{\delta y}\right)^{n-1} ; \quad n < 1 \tag{A.10}$$

Examples of pseudoplastic fluids are: mayonnaise, paper pulp suspensions, high molecular weight polymer solutions, melts of polymers and metals, and suspensions of solids.

**Bingham Fluids:** Bingham fluids (see Curve 4, Figure A–2) require a threshold pressure ($\tau_o$) before they will begin to flow (called the yield value) and are described as follows:

$$\tau_x - \tau_o = \left[\beta\left(\frac{\delta u_x}{\delta y}\right)\right]_f \tag{A.11}$$

The apparent viscosity is defined as:

$$u_a = \frac{[\beta(\delta u_x / \delta y)]_f - \tau_o}{(\delta u_x / \delta y)} \tag{A.12}$$

High concentrations of fine solid (clay and finely ground rocks) suspensions in liquids exhibit the typical behavior of Bingham fluids.

**Rheopectic Fluids:** The behavior of these fluids is far more complex because the apparent viscosity *increases with respect to time* of application of shear stress. Consider Figure A–3a: the arrows on the curve indicate the behavior of apparent viscosity at any point increases with respect to time, as the shear rate changes with respect to time. As before, on Figure A–3a, if a line is drawn from a point on the curve traversed by the apparent viscosity to the origin (dashed line), the slope of the line is the apparent viscosity at that specific

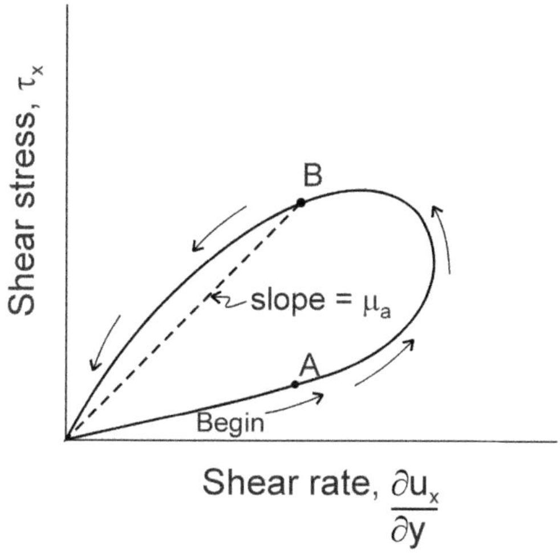


**Figure A–3a** *Rheopectic fluid where the viscosity is time dependent. The viscosity increases with respect to increased shear-rate-time behavior. The slope of the dashed line is the apparent viscosity at any point P.*

*shear rate and time.* However, the curve is dependent on the shear rate time-history; therefore, the curve traversed by the apparent viscosity will not be the same for a different shear rate-time history. Note that points A and B are the stress-shear values at a specific shear rate, but at very different time periods.

Bentonite clay and gypsum suspensions exhibit the properties of rheopectic fluids.

**Thixotropic Fluids:** These fluids are equal in complexity to rheopectic fluids but with one great difference: the apparent viscosity at any point *decreases* with respect to time. The arrows on the curve of Figure A–3b describe a loop that is similar to the one shown for the rheopectic fluid, but the path taken is just the opposite with respect to the shear rate.

Concentrated clay suspensions, such as drilling muds, ink and some paints behave as thixotropic fluids; the apparent viscosity is very high (gel-like) at rest but decreases (thins out) when the fluid is stirred.

Some paints exhibit thixotropic behavior, and in fact it is a very desirable property in paints because the fluid on a stationary brush will have a high viscosity and thus will not tend to flow or drip off of the brush, whereas, when the brush is in motion against a surface, the paint on the brush exhibits a low viscosity that serves to coat the surface evenly and then immediately sets at a high viscosity and will not run down the wall.

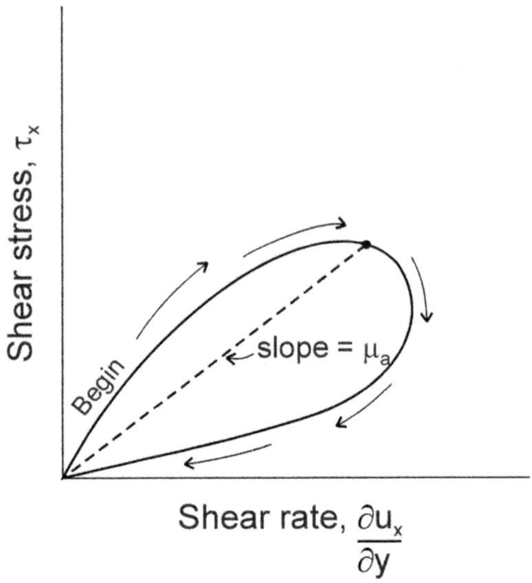


**Figure A–3b** *Thixotropic fluid where the viscosity is time dependent. The viscosity decreases with respect to the increased shear-rate-time behavior. The direction taken by the change of viscosity with respect to time is the reverse of the rheopectic fluid. The slope of the dashed line is the apparent viscosity at any point P.*

**Frac-fluids:** A frac-fluid should have a relatively low viscosity as it is pumped into the formation through the well tubing. The viscosity should be sufficient to entrain the proppant but low enough not to undergo loss of pressure due to turbulence and friction within the tubing. On arrival in the nascent fracture system, the frac-fluid should experience a rapid change into a high viscosity fluid or gel. After completion of the fracture treatment, the frac-fluid viscosity must be reduced during a rest period for temperature increase to reservoir temperature causing a decrease of the frac-fluid viscosity, or for reaction of the viscosity breaker added to the fluid as it was injected.

The frac-fluid will contain 10–80 lbs/1000 gal of a linear or cross-linked polymer to enhance the viscosity in the fracture, or surfactants, which are used to promote high viscosity water external emulsions. Encapsulated breaker compounds that become active during a rest period at the end of the fracture treatment also are used.

APPENDIX B

# Surfactants, Emulsions, Gels, Foams

## B.1 Surfactants

A surfactant is a chemical reagent that will migrate to the surface of a liquid interface between a gas or another immiscible liquid and by its presence at the interface, effectively reduce the surface tension of the liquid. The most common of these are soaps that are composed of two parts: one part is hydrophilic (naturally attracted to water—such as a salt that is easily dissolved by water) and the other is hydrophobic (not attracted to or repelled by water, such as oil). Therefore, at the interface, the hydrophilic portion, or the water-loving part, will be imbedded in the water while the hydrophobic part, or the water-hating part, of the molecule is elevated across the interface between the two fluids and intimately associated with the non-aqueous fluid (see Figure B–1). The interfacial temperature of the oil-water interface may be reduced by 10 to 50 times its normal amount by the presence of soap at the interface.

A soap is the sodium or potassium salt of a fatty acid. The negative carboxylic head $\left(-COO^{-}\right)$ is hydrophilic and the long-hydrocarbon chain made up of linked methyl groups $\left(-CH_2-CH_2-\right)$ is hydrophobic. When soap is dissolved in water, the soap forms a cloudy solution of micelles, which are groups of about 100 soap molecules forming tiny spherical clusters with the ionic heads covering the surface of the sphere while the tails are entangled within the spherical clusters. The micelles are established in the water as stable groups that are hydrogen bonded to surrounding water molecules, and the hydrophobic tails are completely shielded from the water; when oil is present, the micelles may incorporate a droplet of oil within the sphere. Micelles do not begin to form in water until a

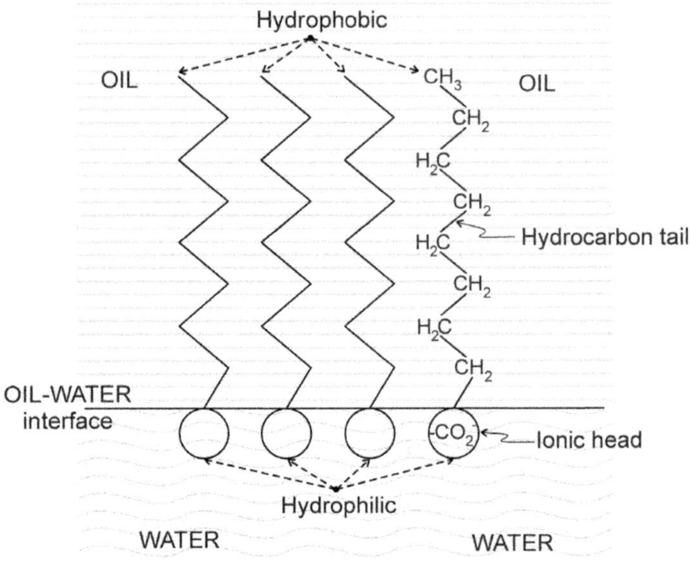


**Figure B–1** *A soap at an oil-water interface has its hydrophobic tails associated with the oil layer on top of the interface and the ionic hydrophilic heads in contact with water below the interface. Sodium ions (not shown) are dissolved in the water around the ionic heads.*

specific concentration of soap (at the prevailing temperature and pressure) is attained. The concentration at which the micelles form is the *critical micelle concentration,* or *CMC,* of the soap at the specific temperature and pressure; soap added to the solution after the CMC is attained will form micelles.

When soap is added to water, the soap molecules will initially congregate at the interface, expelling the hydrophobic parts from contact with water and lowering the interfacial tension. When the CMC is reached, the soap molecules form micelles and decrease the IFT once more.

## B.2 Emulsions

An emulsion is composed of two immiscible liquids, one of which is dispersed completely within a continuous phase. A surface active agent is required to produce a stable emulsion. For an oil-in-water emulsion (where water is the continuous phase) droplets of oil are surrounded by soap molecules whose hydrophobic tails are dissolved in the oil while the hydrophilic heads surround the outer surface, forming a micelle (see Figure B–2). The micelle containing the drop of

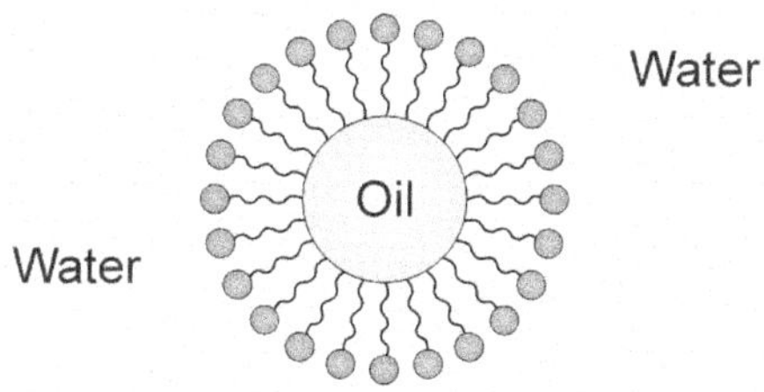


**Figure B–2** *Oil is emulsified in water by forming micelles, which are drops of oil coated with the hyrdophilic, hydrocarbon tails of the soap. The ionic, hydrophilic heads of the soap are associated with water molecules through hydrogen bonding.*

oil is readily suspended within the water because the *hydrophilic* carboxyl groups form hydrogen bonds with the surrounding water molecules. The resulting mixture of one insoluble liquid completely dispersed within another is an *emulsion* (oil has been emulsified by a soap solution). The characteristics of the emulsifying agent (soap) determine whether the emulsion will be an oil-water or a water-in-oil emulsion. If the soap dissolves better in water than it does not in oil, an oil-in-water emulsion forms and vice versa. When the size of the micelles are smaller than the wavelength of light (~0.5 μm), the emulsions appear to be clear; at larger sizes the light rays are scattered, causing the emulsion to appear opaque (or white when all incident light is equally scattered).

Emulsions are inherently unstable and therefore must be agitated in order for the emulsion to form. If left undisturbed, the emulsion will break into its stable two-phase condition. Coalescence, into distinct oil and water phases, occurs if the droplets break and combine into the oil and water phases as the oil floats to the surface. Creaming occurs when the droplets migrate to the surface without breaking, and flocculation occurs if the droplets join to form larger aggregates as they gradually float to the surface. The destabilization of emulsions is accelerated by increased temperature and reaction of the carboxylic heads with water soluble compounds.

## B.3 Foams

If an aqueous solution with surfactant concentration greater than the CMC is agitated by introduction of gas bubbles at the bottom, as the gas bubbles rise to the surface they acquire surfactant molecules at the gas-liquid interface as they ascend to the surface of the liquid. The foam is created at the surface by aggregation of the bubbles. When gas introduction, and turbulence, in the liquid is stopped, the foam at the surface will begin to collapse, or coalesce, to form a continuous oil phase on the surface of the water. The time required for one-half of the liquid making up the foam to break out, at ambient conditions, is the *half-life* of the foam.

Foam is a three-dimensional network of interconnected films (lamellae); the bubbles are generally arranged as three-sided planes radiating at 120° from each connecting point. The liquid-air interface at the surface of the foam is stabilized by a layer of surfactant molecules in the film loosely bound through Van der Waals forces and other molecular electrical attractive forces.

APPENDIX C

# Calculations

## C.1 Subsurface Overburden and Horizontal Stress

The overburden stress (pressure) at any depth is the sum of the stress produced by (1) the average mass of the dry rocks ( $\approx$ 0.54 psi/ft) and (2) the average mass of water containing 80,000 ppm of salts (0.465 psi/ft). Hence, the average overburden stress gradient is $\approx$ 1.0–1.1 psi/ft of depth.

The liquid fluids in the pores of subsurface rocks exert a pressure ($P_p$) that is considered equal to, or in excess of, the hydrostatic gradient. The *effective stress* that governs the failure (fracture) of a rock is the stress of the dry rock matrix minus the pressure exerted by fluids occupying the pores:

$$\sigma_z^* = \sigma_z - P_p \tag{C.1}$$

where

$\sigma^*$ = effective stress

$\sigma$ = natural matrix stress

$P_p$ = pore pressure

The cementation of the rock grains, however, prevents full application of the pore pressure by holding the grains together. A correction factor was introduced by Geerstma (1953) called the poro-elastic constant $(\alpha)$, which is defined as:

$$1.0 - \frac{C_r}{C_b} \tag{C.2}$$

where

$C_r$ = rock matrix material compressibility

$C_b$ = rock bulk compressibility

Thus, the effective vertical stress of a point in the subsurface is:

$$\sigma_z^* = \sigma_z - \alpha P_p \tag{C.3}$$

Several useful relationships are developed using this concept. Eq. (C.10) in the text shows that Poisson's ratio ($\nu$) can be used to define the effective horizontal stress as a function of the effective vertical stress. There are two horizontal stresses, acting at 90° to each other, that are generally considered equal, but in reality one is greater than the other and will be expressed as: $\sigma_x$ = maximum (largest) horizontal stress, and $\sigma_y$ = minimum horizontal stress. Assuming that the horizontal stresses are equal:

$$\sigma_x^* = \sigma_y^* = \left(\frac{\nu}{1-\nu}\right)\sigma_z^* \tag{C.4}$$

This means that the horizontal stresses depend only on the elastic properties of the rock. An average subsurface value for Poisson's ratio is equal to 0.25 and is used to estimate the stress relationships when specific data are not available:»

$$\sigma_x^* = \left(\frac{0.25}{1.0-0.25}\right)\sigma_z^* = \frac{1}{3}\sigma_z^* = \frac{1}{3}\left(\sigma_z - \alpha P_p\right) \tag{C.5}$$

Furthermore, Eq. (C.1) can be used to estimate the changes of effective horizontal stress of layers of different lithology where values of Poisson's ratio are known.

**Example C–1**

Show the difference in the effective stresses of the three types of hard and soft sedimentary formations listed in Table 3.1 of the text for a depth of 5,000 feet:

$$\sigma_z = (1.0 \text{ psi/ft})(5{,}000 \text{ ft}) = 5{,}000 \text{ psi}$$

$$\alpha = 0.7$$

$$P_p = (0.465)(5{,}000) = 2{,}325 \text{ psi}$$

Hard sandstone, $\nu = 0.15$; Effective stress = 595 psi

Soft sandstone, $\nu = 0.20$; Effective stress = 843 psi

Hard dolomite, $\nu = 0.25$; Effective stress = 1,124 psi

Soft dolomite, $\nu = 0.30$; Effective stress = 1,445 psi

Hard limestone, $\nu = 0.23$; Effective stress = 1,007 psi

Soft limestone, $\nu = 0.28$; Effective stress = 1,311 psi

The soft rocks have greater internal stress than those that are hard; and dolomite ($CaMgCO_3$) is a softer rock than limestone ($CaCO_3$).

## C.2 Fracture Initiation Pressure

If the frac-fluid can enter the rock matrix by displacing the formation fluid, the pressure of the frac-fluid ($P_{fr}$) will be sufficient to fracture the rock and hold the fraction open. On the other hand, if the frac-fluid cannot penetrate into the matrix because the pores of the formation are plugged, or the frac-fluid viscosity is too great to allow penetration, or some other circumstance, then an increment of pressure is required to initiate the fracture. The pressure must be increased until it is equal to or greater than the effective least principal stress ($\sigma_y^*$), and using $\nu = 0.25$ for brittle shale from Eq. (C.5):

$$\sigma_y^* = \left(\frac{0.25}{1.0 - 0.25}\right)\sigma_z^* = \frac{1}{3}\sigma_z^* = \frac{1}{3}\left(\sigma_z - \alpha P_p\right) \tag{C.6}$$

and the bottom-hole pressure (at the face of the formation) must be equal to the fracture initiation pressure plus the formation pore pressure:

$$P_{BH} \approx P_{fr} + P_p \approx \frac{1}{3}(\sigma_z - \alpha P_p) + P_P \approx \frac{1}{3}\sigma_z^* + P_P \tag{C.7}$$

For a *horizontal* fracture to occur, the bottom-hole pressure must be greater than the vertical stress and the tensile stress of the formation:

$$P_{BH} = \sigma_z + \sigma_t \tag{C.8}$$

For a *vertical* fracture to develop, the situation is complicated by the circular bore-hole; and the bottom-hole pressure must exceed the minimum horizontal stress as well as the tensile strength of the rock. The derivation for the radial condition reduces to (where the frac-fluid penetrates into the pores):

$$P_{BH} = 3\sigma_x^* - \sigma_y^* + \sigma_t + P_P \tag{C.9}$$

where: $\sigma_x^*$ = maximum horizontal stress; $\sigma_y^*$ = minimum horizontal stress, and $\sigma_t$ = rock tensile stress.

In general, the horizontal stresses are less than the vertical stress and thus a vertical fracture will develop.

The formation closure pressure is the pressure that will close a fracture if it is not propped open by an injected proppant:

$$P_{cl} = \sigma_y + P_P \tag{C.10}$$

## C.3 Fracture Propagation Pressure

In order for a fracture to continue extending away from the bore-hole, the bottom-hole pressure must increase to overcome the increasing resistance to fluid flow imposed by the increasing surface area of the fracture walls as the fracture extends farther into the formation. The pressure of the fluid in the fracture that is in excess of the closure pressure of the formation is labeled the *net fracture pressure*:

$$P_{net} = P_{BH} - P_{cl} \tag{C.11}$$

The pumping pressure at the surface is equal to the net fracture pressure, which is the pumping pressure required to initiate and hold the fractures open against the natural formation closure pressure.

**Example C–2**

Calculate the required surface pressure for the following conditions:

Depth = 7,000 feet

Overburden pressure = 7,000 psi (for a gradient of 1.0 psi/ft of depth)

Pore pressure = 3,220 psi (for a hydraulic pressure gradient of 0.464 psi/ft)

Poro-elastic constant = 0.7

Frac-fluid density = 10lbs/gal = 0.49 psi/ft

Pressure exerted by the column of fluid in the well:
$P_{col} = 7{,}000(0.49\ \text{psi/ft}) = 3426\ \text{psi}$

Effective overburden pressure (see Eq. (C.3)):

$$\sigma_z^* = 7{,}000 - 0.7(3220) = 4746$$

$$P_{BH} = \frac{1}{3}(4746) + (3220) = 4802\,psi$$

$$P_{sur} = P_{BH} - P_{col} = 4{,}802 - 3426 = 1376\,psi$$

The surface pressure also is the minimum pressure that the well casing pipe must withstand.

The stress that the proppant must equal, or exceed, in order to keep the fracture open is the minimum horizontal stress. Near the wellbore, this stress increases during production as the pore pressure decreases near the wellbore, subjecting the proppant to greater stress near the wellbore. Therefore, the proppant strength is specified for the extreme condition of zero pore pressure near the wellbore.

## C.4 Matrix Permeability

Permeability is a rock property that describes the rock's ability to support the flow of fluid when subjected to a pressure gradient. The velocity of the fluid flowing through the rock is a function of the permeability (*K*), the cross-sectional area of the flowing fluid, the viscosity of the fluid, and the difference in pressure across the length of the flow path, expressed as follow in cgs units used in the laboratory:

$$q = \frac{KA}{\mu}\left(\frac{\Delta P}{\Delta L}\right) \quad \left[\frac{cm^3}{s} = \frac{Darcy(cm^2)}{cP}\left(\frac{atm}{cm}\right)\right] \tag{C.12}$$

Permeability is determined in the laboratory under simulated reservoir conditions of overburden stress and reservoir fluid pressure (Donaldson et. al, 1980). Cores taken from the subsurface expand when brought to the surface where the confining pressure is reduced to atmospheric pressure; therefore, the simulated reservoir conditions must be used.

For samples with very low permeability (tight cores), such as those of gas-shale, attempting to obtain steady-state flow (constant flow rate) through the core to measure the permeability is time consuming because the fluid flow rate is so slow. Therefore, a method using transient flow (high pressure in the core is exposed to a low outlet pressure and the change in the rate of fluid flow with respect to time) is used to determine the permeability (Brace, W F et al., 1968).

**Example C–3**

Calculate the permeability of a sandstone core from the follow laboratory data using Eq. (C.12):

$$Q = 0.4\ cm^3/s$$

$$A = 3.0\ cm$$

$$\mu = 0.9\ cP\ at\ 77^{\circ}F$$

$$\Delta P = 2.5\ atm$$

$$L = 2.5\ cm$$

$$K = \frac{0.4\,x\,0.9\,x\,2.0}{3.0\,x\,2.5} = 0.096\ Darcy$$

# C.5 Fracture Size

There are two prominent theories of fracture design: the Perkin-Kern-Nordren model (PKN model) assumes the width of the fracture is a function of the height and that the fracture is elliptical in shape (Perkins and Kern, 1961; Nordgren, 1972). The other is the Khristianovic-Geertsma-de Klerk model (KGD model) that is based on the assumption that the width of the fracture is a function of its length and that the fracture is rectangular in shape (Khristianovic, 1955; Geertsma and de Klerk, 1969). Neither method has been proven superior to the other, and the use of these models is qualitative rather than quantitative. Newer, more sophisticated models have been developed using partial differential equations that must be solved by a computer. These are beyond the scope of this book.

The KPN model assumes that the fracture height stops at the upper and lower boundaries of the formation, thus if the fracture has a fixed height which is independent of the length of the fracture. Other assumptions are that the fracture pressure is constant within a given position of the fracture. The fracture develops with an elliptical shape and has an average width, and Young's modulus and Poisson's ratio are constant rock properties. These relationships translate into a mathematical expression the yields the fracture geometry for a fracture that extends on both sides of the wellbore:

***KPN Model***

$$W = (0.45)\left[\frac{(1-\nu)q^2\mu}{GH}\right]^{1/5}\left(t^{1/5}\right) \qquad \text{(C.13)}$$

$$L = (1.89)\left[\frac{Gq^3}{(1-\nu)\mu H^4}\right]^{1/5}\left(t^{4/5}\right) \qquad \text{(C.14)}$$

The GDK model yields larger fracture widths than the KPN method and in cases where the actual widths of fractures have been measured, the widths calculated by the GDK model have been found to be closer to the widths of actual fractures. The height of the fracture is assumed to be constant throughout the length, the injection rate is constant.

### *GDK Model*

$$W = (1.32)\left[\frac{(1-\nu)q^3\mu}{GH^3}\right]^{1/6}\left(t^{1/3}\right) \tag{C.15}$$

$$L = (0.48)\left[\frac{Gq^3}{(1-\nu)\mu H^3}\right]^{1/6}\left(t^{2/3}\right) \tag{C.16}$$

**Example C–4**

Calculate the fracture width and length using the PKN and GDK models and compare the results.

$G = 1.3 \times 10^6$ psi $= 9 \times 10^6$ kPa

$\nu = 0.24$

$H = 80$ feet $= 24.4$ m

$q = 10$ bbl/min $= 1.6$ m$^3$/min

$t =$ cP $= 1.67 \times 10^6$ kPa-min

The table below shows Young's modulus and shear stress for shale. The first three points indicate brittle shale, with $\nu = 0.25$ as the division between brittle and ductile shale; ductile shale values are the last four in the table, where

$$G = \frac{E}{2(1+\nu)}$$

| $E \times 10^6$ | $G \times 10^6$ | n | E/ν |
|---|---|---|---|
| 2.8 | 1.2 | 0.21 | 13.3 |
| 3.2 | 1.3 | 0.24 | 13.3 |
| 3.7 | 1.5 | 0.25 | 14.8 |
| 4.1 | 1.6 | 0.27 | 15.2 |
| 4.4 | 1.7 | 0.29 | 15.2 |
| 4.8 | 1.8 | 0.32 | 15.0 |

***KPN Model***

$$W = (0.48)\left[\frac{(0.76)(2.56)(1.67\,x10^{6}}{\left(9\,x10^{6}\right)(24.4)}\right]^{1/6}\left(150^{1/5}\right) = 8.79\,x10^{-3}m = 0.346\ inch$$

$$L = (0.45)\left[\frac{\left(9\,x10^{6}\right)\left(1.6^{3}\right)}{(0.76)\left(1.67\,x10^{-6}\right)\left(24.4^{4}\right)}\right]^{1/5}\left(150^{4/5}\right) = 946\ m = 3{,}104\ feet$$

***GDK Model***

$$W = (1.32)\left[\frac{(0.76)(1.6)\left(1.67\,x10^{-6}\right)}{\left(9\,x10^{6}\right)\left(14.53\ x10^{3}\right)}\right]^{1/6}\left(150^{1/3}\right) = 16\,x10^{-3}\ m = 0.63\ inch$$

$$L = (0.48)\left[\frac{\left(9\,x10^{6}\right)(4.1)}{(0.76)\left(1.67\,x10^{-6}\right)\left(1.45\,x10^{4}\right)}\right]^{1/6}\left(150^{2/3}\right) = 479\ m = 1{,}571\ feet$$

## C.6 Rock Matrix Mechanical Properties

The in situ rock conditions of stress, porosity, fluid saturations, etc., are measured using instruments that are lowered into the wells. Before the advent of computers, the logs were interpreted by experienced log analysts. The log analyst today is assisted by computers that can produce refined evaluations of subsurface petrophysical properties, many of which are calibrated, or verified, by analyses of cores in the laboratory. Explanations of the most important logs are presented with examples using realistic data.

The acoustic log is used to determine porosity and the rock mechanical properties. The instrument is designed to emit acoustic waves at about 20,000 Hertz in pulses at 15 times per second. The interval of time between emission of the pulse and its arrival at receivers spaced at specific distances from the source is recorded. The transit time, μs/ft or velocity, ft/s, is a function of the petrophysical properties through which it travels and is used to calculate the properties.

Porosity is determined from Wyllie's time-average formula:

$$\phi = \frac{t - t_m}{t_f - t_m} \quad \text{(C.17)}$$

Where $t$ = acoustic time recorded by the log, μs/ft; $t_m$ = acoustic time of the type of rock such as sandstone, limestone, etc., determined in the laboratory; $t_f$ = acoustic time of the particular formation fluid occupying the pores, also determined in the laboratory: water with 20% sodium chloride at ambient temperature (77°F) is 189 μs/ft. Unconsolidated silica sand ranges from 58 to 62 μs/ft; dry consolidated sandstone is 53; limestone is 48, dolomite is 43; and shale ranges from 60 to 160 μs/ft.

The overburden pressure increases the velocity of acoustic waves (decreases the transit time); therefore, a correction factor is used for log interpretation where laboratory core analyses are not available for correlation.

$$C_p = \frac{t_{sh} \, x \, c}{100} \quad \text{(C.18)}$$

Where $t_{sh}$ = specific acoustic transit time in the shale formation; $c$ = compaction value characteristic of the shale that varies between 0.8 and 1.2; and 100 is the average transit time in compacted shales.

**Example C–5**

Calculate the porosity of a shale formation whose specific transit time *(t)* is 115 μs/ft; matrix time ($t_{ma}$) is 110 μs/ft; fluid time ($t_f$) is 189 μs/ft and the compaction coefficient (*c*) is 0.8:

$$\phi = \frac{t - t_{ma}}{t_f - t_{ma}} x \frac{t \, x \, c}{100} \qquad \phi = \frac{115 - 110}{189 - 110} x \frac{115 x 0.8}{100} = 0.058 = 5.8\%$$

The travel velocity of the acoustic wave through subsurface rocks is also used to determine the stress properties of the formation of interest. The sound wave separates into three signals as it travels through the rock. The first signal to arrive at the detector is the compression wave which is transmitted through solids, gases and liquids because they resist the compression of their atomic structure. The compression wave propagates through the medium by a series of compressions and rarefactions as it travels away from the source; in other words, the rock particles are moved back and forth as the wave travels forward. The compression wave's velocity is a function of: (1) the density of the material, (2) the bulk modulus and (3) the shear modulus; the relationships are expressed mathematically as:

$$u_c = \left( \frac{K_B + \frac{4}{3}G}{\rho} \right)^{\frac{1}{2}} \quad \mu s/ft \tag{C.19}$$

The shear wave travels at about half the speed of the compression wave by displacing particles in the direction perpendicular to the direction of travel; that is, the particles are moved up and down as the wave travels forward. Solids offer resistance to this movement, or oppose the shearing force generated by the wave, but gases and liquids are not rigid, and, hence, shear waves cannot be propagated through them. The shear wave is less complex being a function only of the density of the material and the shear modulus, which is expressed as:

$$u_s = \left( \frac{G}{\rho} \right)^{\frac{1}{2}} \quad \mu s/ft \tag{C.20}$$

Poisson's ratio ($\nu$) and the shear modulus ($G$) are calculated from the transit times (as defined above for the porosity determination) and density of the material. Using the values of the shear modulus, Young's modulus is calculated from the Poisson's ratio and the shear modulus, as expressed in Eq. (C.21).

$$\nu = \frac{0.5\left(\frac{t_s}{t_c}\right)^2 - 1}{\left(\frac{t_s}{t_c}\right)^2 - 1} \tag{C.21}$$

$$G = \frac{1.34(10^{10})\rho}{t_s^2} \quad psi$$

$$E = 2G(1-\nu) \quad psi$$

**Example C–6**

Calculate the stress properties of Gulf Coast shale lying at a depth of 8,000 ft. The compression wave transit time $(t_c)$ is 115 μs/ft; the shear wave transit time $(t_s)$ is 201 μs/ft; and the bulk density of the shale is 2.55. The ratio of shear to compression times, $t_s/t_c = 1.75$.

$$\nu = \frac{0.5(1.75)^2 - 1}{(1.75)^2 - 1} = 0.256$$

$$G = \frac{(1.34x10^{10})(2.55)}{(201)^2} = 0.843x10^6 \quad psi$$

$$E = 2(0.843x10^6)(1+0.256) = 2.63x10^6 \quad psi$$

The ratios of shear to compression transit times for other materials are approximately: 1.78 for sandstone containing water, 1.60 for sandstone containing gas, 1.80 for dolomite, and 1.90 for limestone.

Archie's Equation (Eq. (2.2) without the $V_{sh}$ term) is more or less a standard for calculation of water saturation. But its accuracy, when applied to gas-shale is questionable because there are so many variable parameters that are taken into account. Variable amounts of conductive minerals in the shale, such as pyrite, lower the shale resistivity;

the salinity of water in the pores varies over short distances; the different types of kerogen increase the shale resistivity by different amounts; and the values of the constants *a, m,* and *n* to be used in Archie's equation cannot be standardized. The results of the equation must be correlated with laboratory data, but even so, the changing properties of the gas-shale both vertically and horizontally produce uncertainties. Nevertheless it is used as an indicator of the hydrocarbon content (or total organic carbon) of the shale when correlated to laboratory data.

**Example C–7**

Calculate the water saturation and TOC, using Archie's equation, assuming the following conditions: $a = 1.0$; $m = 1.92$; porosity($\phi$) = 8%; water resistivity ($R_w$) = 0.05; $R_t = 40$:

$$S_w^n = \left(\frac{1.0}{\phi^{1.92}}\right)\left(\frac{Rw}{R_t}\right) = S_w^2 = \left[\frac{1.0}{(\phi)^{1.92}}\right]\left(\frac{0.05}{40.0}\right)^{\frac{1}{2}} = 0.398$$

$$\textit{Hydrocarbon Saturation} = 1.0 - S_w = 1.0 - 0.398 = 0.602 \textit{ or } 60.2\%$$

## C.7 Summary

The initial phase of characterization of a gas-shale reservoir is summarized as follows:

1. Identify the shale and its areal, thickness, and depth variation using a surface seismic survey.
2. Drill a vertical hole, taking samples of the target shale for laboratory analyses of: mineralogy, vitrinite reflectant, porosity, water saturation, total organic carbon content (TOC), water salinity and saturation, gas saturation and composition, Young's modulus and Poisson's ratio.
3. Run wire-line well-logs: electric-log, porosity, acoustic travel time, total organic carbon, gamma-ray and spectral-gamma

ray, caliper, neutron and pulsed neutron, NMR (nuclear magnetic resonance).

4. Compare and calibrate the well-logs with the laboratory analyses.

5. Conduct calculations using computer program algorithms and imaging to select intervals for completion, if the vertical well is to be used for production.

6. Drill lateral (horizontal) wells radially into the shale reservoir using Logging-While-Drilling for selection of the best zones for production stimulation (hydraulic fracturing).

7. Case and perforate the horizontal well for multiple stage fracture treatment.

# Glossary

**Basin** A large bowl-shaped indentation in the earth's surface; since the sides slope toward the middle of the basin, organic and inorganic particles, transported by wind and water, migrate to the center, hence, the structures of shale and rock in the center are the youngest structures. The oldest are around the rim of the bowl.

**Casing** Steel pipe placed in a well and generally cemented by squeezing cement down the center of the casing and back to the surface in the annulus between the casing and the borehole.

**Clay** Often a major component of soils. It is plastic when wet, but hard and brittle when dry and compacted, and typically composed of hydrous aluminum silicates.

**Completion** After the well is cased, cemented, and fractured, cleaning out the well to remove impediments to production is done and equipment required for production is installed, such as pumps, pipes for transport and other final operations to place the well on production status.

**Conventional Reservoir** Reservoirs that have high permeabilities and are located in sandstone and carbonate reservoirs. Unconventional Reservoirs exhibit very low permeabilities (< 0.01 md) and are located in shale beds or very low permeability sedimentary rocks.

**Drilling mud** A mixture of bentonite clay and some compounds (such as potassium chlorite and surfactants) that is used to lubricate the drill-bit and drill-stem, and transport the rock fragments (cuttings) produced by the bit to the surface.

**Feldspar (Orthoclase)** Potassium Aluminum Silicate KAlSi3O8; Plagioclase: Calcium Aluminum Silicate CaAl2Si2O8; Albite: Sodium Aluminum Silicate NaAlSi3O8)

**Flow line** A pipe connecting the well to processing equipment (water separator, emulsion breaker, temporary accumulation, etc.).

**Foot print** The surface area required for installation of a drilling rig and its associated equipment and vehicle parking.

**Formation** A subsurface geologic structure composed of specific, mostly isotropic, materials (sand, clay, limestone, etc.) that is easily recognized throughout its extent as distinguished from other structures around it.

**Frac-fluid** An aqueous, or oil based, mixture containing small amounts of special chemical compounds that is used for hydraulic fracture stimulation of gas and oil production from a subsurface hydrocarbon reservoir.

**Geophone** An instrument for detecting vibrations traveling through rock, soil, or ice.

**Hydrostatic pressure** Pressure below the surface of a column of liquid, which is a function of the fluid density, depth, and temperature.

**Incompatible fluids** Fluids that promote precipitates when mixed.

**Laterals** Horizontal wells that extend laterally (horizontally) away from the bottom of a vertical well.

**Leak off** Loss of frac-fluid from fractures to the formation matrix by displacement of reservoir fluids.

**Net gas play** The thickness of a gas producing zone of a reservoir; for example, a formation may be 100 feet thick, but only a top portion (say 20 feet of the top of the reservoir) actually produced gas.

**Original gas in place** The best estimate of the total volume of gas contained within the entire shale formation. This estimate may change during the active life of the reservoir if new information becomes available from analyses of production records, laboratory data, or new developments of technology.

**Product** Chemicals used and waste generated by hydraulic fracturing.

**Proved reserves** Recoverable, mobile gas in place that has been proven to exist by definitive laboratory or production well tests.

**Resistivity** Resistivity is the resistance (I/E = current (amperes)/voltage = ohms) multiplied by the cross section area presented to the flow of current, and divided by the length through which the current flows: $R = (I/E)(m^2/m)$ = ohm-meters.

**Seismic** Earth vibrations cause by earthquakes, explosives, mechanical impacts, the fracture of rocks, etc.

**Shale** Formed by consolidation/compaction of clay, mud, and silt. Contains a substantial amount of clay...generally more than 60%.

**Silica flour** Pure silica sand that has been ground into a very fine powder.

**Standard conditions** 60°F and one atmosphere (14.7 psi) of pressure.

**Tight gas formation** A very low permeability formation containing a commercial amount of natural gas. The gas cannot be produced unless the well is stimulated by hydraulic fracturing.

**Volumetric symbols** Mscf one thousand (103) standard cubic feet; MMscf one million (106) standard cubic feet; Bscf one billion (109) standard cubic feet; Tscf one trillion (1012) standard cubic feet (enough gas to supply approximately 15 million homes per year).

# References


Allen, T O and Robert, A P, 1994. *Production Operations, V. 2. PennWell Books,* Tulsa: 142–150.

American Petroleum Institute, 2010. Hydraulic Fracturing: Unlocking America's Natural Gas Resources. www.scribd.com/doc/34640953/ Hydraulic-Fracturing-Unlocking-America%E@%80%99s-Natural-Gas-Resources.

Archie, G E, 1942. The electrical resistivity of logs as an aid in determining some reservoir characteristics. *Trans. AIME,* v. 146.

Asgharian, B; Price, O T; et al., 2012. A lung dosimetry model and vapor uptake and tissue disposition. *Inhalation Toxicology,* v. 24 (3), Taylor and Francis Group Publication, London, England: 182–193.

Bayazeed, A F and Donaldson, E C, 1970. *Subsurface Disposal of Pickle Liquor.* US Bureau of Mines Report of Investigations 7804, *National Technical Information Services, Springfield, VA*: 32 pp.

Bayazeed, A F and Donaldson, E C, 1971. Deep-well disposal of steel pickling acid. SPE 3625, *46th Annual Technical Conf., New Orleans, LA* (October), 8 pp.

Biot, M A, 1955. Theory of elasticity and consolidation for a porous anisotropic solid. *J. Applied Physics,* v. 26 (Feb): 182–197.

Brace, W F; Walsh, J B and Francos, W T, 1068. Permeability of granite under high pressure. *J. Geophy Res.,* v. 73: 2225–2336.

Breit, G N, 2002. Produced waters database. US Geol Survey. www.energy.cr.usgs.gove.prov/prodwat/index.htm.

Chilingarian, G V; Donaldson, E C and Yen, T F, 1995. *Subsidence due to Fluid Withdrawal.* Elsevier Publishing Co, Amsterdam, New York: 498 pp.

Coats, B, 2012. Microseismic reservoir pressure monitoring advances hydraulic stimulation effectiveness in Bakken shale. *World Oil,* Fracturing Technology 2012—Special supplement, March, 15–17.

Coats, G R and Denoo, S A, 1961. Mechanical properties program using borehole analysis and Mohr's circle. *SPWLA Twenty-second Annual Logging Sym.*, June: 1–15.

Cook, N G W, 1967. The design of underground excavations. *Proc. 8th Sym. On rock Mechanics.* Minnesota University: 167–193.

Daneshy, A., 2011. Hydraulic fracturing of horizontal wells: issues and insights. SPE 140134, *Hydraulic Fracturing Technology Conf. and Exhibition, The Woodlands, TX,* (Jan).

Donaldson, E C 1964. *Subsurface Disposal of Industrial Wastes in the United States.* US Bureau of Mines Information Circular 8212. *National Technical Information Service, Springfield, VA*: 34 pp.

Donaldson, E C, 1974. Injection wells and operations today. Memoir 18. *Am. Asso. Pet. Geol.,* 24 pp.

Donaldson, E C and Alam, W, 2008. *Wettability*. Gulf Pub. Co., Houston, TX: 336 pp.

Donaldson, E C; Kendall, R F; Pavelka, E A and Crocker, M E, 1980. Equipment and procedures for fluid flow and wettability tests of geologic materials. IC-79/5. *National Technical Information Service, Springfield, VA*: 40 pp.

Donaldson, E C, Thomas, R D, and Johnston, K H, 1974. *Subsurface Waste Injection in the United States—Fifteen Case Histories.* US Bureau of Mines Information Circular 8636, *National Technical Information Service, Springfield, VA:* 72 pp.

Dmour, H N and Eissa, M S, 2008. Pre-post test data analysis for hydraulically fractured vertical tight gas well-field case study. *Oil and Gas Business.* www.ogbus.ru/eng/authors/Dmour/Dmour_1.pdf.

Economides, M J and Nolte, K G, 1987. *Reservoir Stimulation.* Schlumberger Educational Services, Houston, TX.

Gaudlip, A W; Paugh, L O and Hayes, T D, 2008. Marcellus Shale water management challenges in Pennsylvania, SPE 229898. *Shale Gas Production Conf.*, Fort Worth, TX (Nov 16–18).

Glaydos, J S and Harris P C, 1980. Forma fracturing: theories, procedures and results. SPE 8961, SPE/DOE Sym. *On Unconventional Gas Recovery,* Pittsburg, PA, May.

Geerstma, J., 1953. The effect of fluid pressure decline on volumetric changes of porous rocks. *Trans. AIME,*: 331–340.

Geerstma, J and de Klerk, F, 1969. A rapid method of predicting width and extent of hydraulically induced fractures. *SPE J. Petrol. Tech.* (Dec): 1571–1581.

Gidley, J L; Holditch, S A, Nierode, D E and Veach Jr, R W, 1989. *Recent Advances in Hydraulic Fracturing.* Monograph v. 12, *Soc. Petrol. Engr:* 452 pp.

Grebe, J and Stoesser, S M, 1935. Increasing crude production from established fields. *World Petroleum,* v. 6 (8), August: 373–482

Grieser, B and Bray, J, 2007. Identification of production potential in unconventional reservoirs. SPE 106623, *SPE Production Operations Symp.,* Oklahoma City, OK, (31 Mar to 3 Apr).

Gupta, D V S, 2009. Unconventional fracturing fluids for tight gas reservoirs. SPE 119424. *Hydraulic Fracturing Tech. Conf.,* The Woodlands, TX, Jan 2009.

Halliburton Co.2012.HydraulicFracturing. www.halliburton.com/public/projects/pubsdata/Hydraulic.Fracturing/index.html?SRC=MP.

Hays, T, 2009. Sampling and analysis of water streams associated with the development of the Marcellus Shale gas. Final Report prepared for Marcellus Shale Coalition. *Gas Technical Institute,* Des Plains, IL, 1700 S Mount Prospect Rd, Des Plaines, IL 60018: 249 pp.

Hlidek, B T and Rieb, B, 2011. Fracture stimulation treatment best practices in the Bakken Oil Shale. SPE 140252, *Hydraulic Fracturing Tech. Conf.,* The Woodlands, TX, (Jan): 10 pp.

Holcomb, D S; Callaway, E. and Curry, L.L., 1980. Foamed hydrocarbon stimulation of water sensitive formations. SPE 9033, *SPE Rocky Mt. Regional Meeting,* Casper, WY, May.

Hubbert, M K and Willis, D G, 1957. Mechanics of hydraulic fracturing. *Trans. AIME* 210: 153–166.

Performance Indices, 2012. *SPE J. Petr. Tech,* (May): p. 6.

Kalish, P J, 1964. The effect of bacteria on sandstone permeability. *J. Petrol. Tech.,* July: 805–814.

Keller, K A and Snyder, C A, 1986. Mice exposed in utero to low concentrations of benzene exhibit enduring changes in their colony forming hematopoietic cells. *Toxicology,* v. 42. Elsevier Pub. Co: 171–181.

Khiristianovic, S A and Zheltov, Y P, 1955. Formation of vertical fractures by means of highly viscous liquids. *Proceeding Fourth World Congress,* Rome.

King, G E, 2010. Thirty years of gas shale fracturing: What have we learned? SPE 133456. *SPE Ann. Tech. Conf.,* Florence, Italy, (Sep).

Ksoy, M, 1989. Hematotoxicity and carcinogenicity of benzene. *Environmental Health Perspective,* v. 82, North Carolina: 97–108.

Kundert, D and Mullen, M, 2009. Proper evaluation of shale gas reservoirs leads to a more effective hydraulic-fracture stimulation. SPE 123586. *SPE Rocky Mountain Petrol. Tech. Conf.,* Denver, CO, April.

Lyons, W C and Plisga, G J, 2005. *Standard Handbook of Petroleum and Natural Gas Engineering.* Chapter 6. Elsevier Pub. Co., New York, Amsterdam.

Mullen, M; Roundtree, R; Baree, R and Turk,G, 2007. A composite determination of mechanical rock properties for stimulation design. SPE 108139, *Rocky Mt. Oil & Gas Tech. Sym.,* Denver, CO, 16–18 Apr.

Nordgren, R P, 1972. Propagation of a vertical hydraulic fracture. *SPE Journal* (Aug): 306.

Penny, G S, 1985. Control and modeling of fluid leak-off during hydraulic fracturing. *SPE J. Petrol. Tech.,* June: 1071.

Perkins, K T and Kern, L R, 1961. Widths of hydraulic fractures. *SPE J. Petrol. Tech.*: 937–949.

Reis, J C, 1996. *Environmental Control in Petroleum Engineering.* Gulf Pub. Co., Houston, TX: 274 pp.

Rickmon, R; Mullen, M; Petre, E; Grieser, B and Kundert, D, 2008. A practical use of shale petrophysics for stimulation design optimization: all shale plays are not clones of the Bernett shale. *SPE 115258, Ann. Tech. Conf. and Exhibition,* Denver, CO, 21–24 Sep.

Rothman, N; Li, G L; Dosemeci, M, Bechtold, G E, Marti, Y Z, Wang, M; Linet, L Q, Xi, W; Lu, M T; Smith, N; Titendo-Holland, L °; Zhang, W; Blot; S N, and Hayes, R B, 1996. Hematotoxicity among Chinese workers heavily exposed to benzene. *Am. J. Ind. Med.,* v. 29, Wiley Periodicals, Wilmington, DE: 236–246.

Rowan, E I; Engle, M A; Kirby, C S and Kraemer, T F, 2011. Radium content of oil-and-gas-field produced water in the Northern Appalachian Basin (USA): Summary and Discussion Data. Scientific Investigation Report 2011-5135, US Geol. Survey, Reston, VA. 31 pp. http://pubs.usgs.gov/sir/2011/5135/.

Sandungaray, P and Palisch, T, 2012. Hydraulic fracture optimization in unconventional reservoirs. *World Oil,* Fracturing Technology 2012—Special Supplement, March, 7–13.

Satterfield, J, 2008. Managing water resource's challenges in selected natural gas shale plays. *Proc. 25th Ground Water Protection Council Annual Forum,* (Abstract 46). The Ground Water Protection Council 13308 N. MacArthur, Oklahoma City, OK 73142.

Stegent, N; Ingram, S and Callard, J, 2011. Hydraulic fracture stimulation design considerations and production analysis. SPE 139981, *Hydraulic Fracturing Tech. Conf.* The Wooklands, TX (Jan): 10 pp.

Sun, H Wood, B; Stevens, D; Cutler, J; and Qi, Q; 2011. A nondamaging friction reducer for slickwater frac applications. SPE 139480. *Hydraulic Fracturing Tech. Conf.,* The Woodlands, TX, January.

Texas Railroad Commission, 2011. Water use in the Barnett shale. (Jan). www.bseec.org/content/water-use-barnett-shale-gas-wells.

Texas Waste Development Board, 2007. Northern Trinity/Woodbine aquifer groundwater availability model: Assessment of groundwater use in the Northern Trinity Aquifer due to urban growth and Barnett Shale Development. http://rio.twdb.state.tx.us/RWPG/rpgm_rpts/0604830613_BarnetShale.pdf.

Tiab, D and Donaldson, E C, 2004. *Petrophysics (2nd Ed.—2004; 3rd Ed.—2012) Elsevier Science Publishers*, Amsterdam, New York, 889 pp.

US Dept. of Energy, 2007. World proved reserves of oil and natural gas. US EnergyAdm. ww.eia.doe.gov/emeu/international/reserves.html.retrived.

US Dept. of Energy, 2008. Annual Energy Outlook 2008 with projections to 2030. US Energy Adm. www.eia.doe.gov/oiaf/aeo/pdf/0383(2008).pdf. June 2008.

US Dept. of Energy, 2009. International Energy Statistics 2009. US Energy Info. Adm. ww.eia.gov/cfapps/ipdbproject/IEDIndex3.cfm?tid=5&pid=57&aid=6

US Dept. of Energy, 2009. Modern shale gas development in the United States—A Primer (Apr). *Office of Fossil Energy,* National Energy Tech. Lab. (Apr): 96 pp.

US Dept. of Energy, 2012. Annual Energy Outlook. US Energy Adm. www.eia.gov/forecasts/aeo/.

US Geological Survey, 2010. Digital Data Services. www.eagleford.com/projects/images_matthews/diagrammatic-cross-section.gif

US Energy Information Agency, 2010. Map of Eagle Ford Shale Play, Western Gulf Basin, South Texas. www.eia.gov/oil_gas/rpd/shaleusa9.pdf

US Environmental Protection Agency, 2002. A review of the reference dose and reference concentration processes. EPA/630/°-02/002F, (Dec). National Technical Information Service, Alexandria, VA.

US Environmental Protection Agency, 2003. Toxicological Review of benzene. EPA/635/R-02/001F (Dec), National Technical Information Service, Alexandria, VA.

US Environmental Protection Agency, 2005. Guidelines for carcinogen risk assessment. EPA/630/P-03/00F, (Mar). Technical Information Service, Alexandria, VA.

US Environmental Protection Agency, 2005. Supplemental guidance for assessing susceptibility from early-life exposure to carcinogens. EPA/630/R-03/003F, (Mar). ). Technical Information Service, Alexandria, VA.

US Environmental Protection Agency, 2006. A framework for assessing health risks and environmental exposure to children (Final). EPA/600/R-05/093F (Sep). Technical Information Service, Alexandria, VA.

US Environmental Protection Agency, 2011. Recommended use of body weight ¾ as the default method in derivation of the oral reference dose. EPA/100/R11/000/ Technical Information Service, Alexandria, VA.

US Environmental Protection Agency, 2011. Hydraulic Fracturing Study Plan. (Nov) EPA/6000/R-121/122. www.epa.gov/hfstudy/HF_Study_Plan_110211_FINAL_508.pdf.

Veatch, R W., Jr; Zissis, A M and Fast, C R In: Gidley, et al. (Editors), 1989. *Recent Advances in Hydraulic Fracturing*. SPE Monograph 12: 452 pp.

Veley, C D., 1969. How hydrolizable metal ions react with clay to control formation water sensitivity. *J. Petro. Tech*, September: 111–1118.

Vidic, R D, 2010. Sustainable water management for Marcellus Shale development. Dept. Civil and Environmental Eng., U. of Pittsburgh. www.temple.edu/environment/NRDP+pics/shale/presentations_TUsummit/Vidic-Temple-2010.pdf.

Waxman, H A; Markey, E J and DeGette, D, 2011. Chemicals used in hydraulic fracturing. US House of Representative Committee on Energy and Commerce (Apr), http://democrats.energycommerce.house.gov/sites/default/files/documents/Hydraulic%20Fracturing%20Report%204.18.11.pdf.

William, S, 2012. Discovering shale gas: an investor guide to hydraulic fracturing. *The sustainable Investment Institute (Si2).* IRRC Institute, New York.

Winter, T C; Harvey, J W; Franke, Q L and Alley, W M, 1998. Ground water and surface water. A single resource. US Geol. Survey Circular 1139. Rocky Mt. Mapping Center, Box 25046, Denver Federal Center, Denver, CO 90225.

Wyllie, M R J; Gregory, A R and Gardner, L W, 1956. Elastic wave velocities in heterogeneous porous media. *Geophysics,* Jan: p. 21

Zielinski, R A and Oton, J K, 1999. Naturally occurring radioactive material 9NORM) in produced water of oil-field equipment—An issue for the energy industry. US Geol. Survey. http://pbus.usgs.gov/fs/fs-0142-99/fs-0142-99.pdf.

# Index

CPI Antony Rowe
Chippenham, UK
2019-02-20 22:57